Corrosion Protection of Metals by Intrinsically Conducting Polymers

Corrosion Protection of Metals by Intrinsically Conducting Polymers

Pravin P. Deshpande
Dimitra Sazou

CRC Press
Taylor & Francis Group
Boca Raton London New York

CRC Press is an imprint of the
Taylor & Francis Group, an **informa** business

CRC Press
Taylor & Francis Group
6000 Broken Sound Parkway NW, Suite 300
Boca Raton, FL 33487-2742

First issued in paperback 2021

© 2016 by Taylor & Francis Group, LLC
CRC Press is an imprint of Taylor & Francis Group, an Informa business

No claim to original U.S. Government works

Version Date: 20160115

ISBN-13: 978-0-367-78326-6 (pbk)
ISBN-13: 978-1-4987-0692-6 (hbk)

Visit the Taylor & Francis Web site at
http://www.taylorandfrancis.com

and the CRC Press Web site at
http://www.crcpress.com

Contents

Preface

This book is devoted to the perspective of the intrinsically conducting polymers (CPs) in metal anticorrosion technology. The application of CPs as protective coatings on the surfaces of active metals was not a straightforward process. However, if one looks back 20–30 years, definite progress has been made in understanding the mechanisms by which CPs provide protection to metals in corrosive environments, building up strategies to improve CP anticorrosion performance, developing techniques to evaluate protective efficiency, and devising new formulations of CP-based coatings, appropriate to be applied on surfaces of a variety of metals and alloys with the ultimate goal of long-term protection of metal substrate by retaining and exploiting simultaneously the self-healing action of CPs. The self-healing property of CPs keeps the interest alive in the field making CP-based coatings as the most suitable alternates of coatings based on the environmentally hazardous hexavalent chromium as environmental restrictions are continuously enhanced worldwide.

It was found timely to try to present an overview of the present status of concepts and methods, which should be essential to pursue the activity in the field of corrosion protection of metals by intrinsically CPs. This presentation is in a sense a complement to recent reviews in the field, but with emphasis on basic aspects of the corrosion problem, CP electronic/ionic properties, and the role of CPs to act as *green* inhibiting coatings against metal corrosion. Certainly, the book includes only a number of topics, which had been selected according to our own interests and fields of expertise. However, the reader will easily find in cited references other works where different topics are discussed, sometimes with other points of view.

The first part of the book gives an introduction to basic concepts related with CP-based coatings utilized in metal protection against corrosion. As the topic is interdisciplinary, properties of CPs with emphasis on their electronic conductivity, metal corrosion processes, and the promising role of CPs in preventing and controlling the metal corrosion are briefly outlined. The second part provides fundamentals of metal corrosion as an electrochemical process. Thermodynamic and kinetic aspects are considered. The third part deals with corrosion prevention mechanisms in terms of which the CP protection capability can be understood. Many aspects of the preparation of CP-based coatings are treated in the fourth part. Examples of a variety of effective formulations, already established through many years of intensive research, are presented. The fifth part describes electrochemical techniques by which anticorrosion properties of CP-based coatings can be evaluated. Laboratory, pilot plant, or field corrosion tests look for passive metal surfaces with neglected corrosion rates and noble equilibrium potential of the M|CP interface. Promising results have been a key to progress and a motive for developing new approaches to improve the CP coating protective performance. Problems were sometimes recognized to arise upon a long-term exposure of M|CP coating systems to corrosive environments. An attempt to outline the most important of these problems, which have induced new strategies for improving the protective performance of CP coatings, is made in the sixth part.

In the course of writing this book, we received valuable feedback from colleagues and undergraduate/graduate students. We will avoid to cite them individually for fear of forgetting one of them and of disturbing our friendship. In any case, their names appear all over the related references and citations. D. Sazou would nevertheless like to thank especially Prof. Michael Pagitsas, whose expertise was in the field of theoretical physical chemistry, for his constant encouragement and support for the intense collaboration they had on several topics of metal corrosion science and, in particular for making bridges between their approaches.

P.P. Deshpande is much indebted to Prof. B.B. Ahuja, Officiating Director, College of Engineering Pune (COEP), and Prof. A.D. Sahasrabudhe, Chairman, All India Council of Technical Education (AICTE), New Delhi, for their constant support; Prof. N.B. Dhokey, Head, Department of Metallurgy and Materials Science, College of Engineering Pune (COEP), and Prof. S.T. Vagge for extending facilities for the experimental work and valuable discussions; Prof. S.U. Pathak and Prof. D.R. Peshwe, Department of Metallurgical and Materials Engineering, Visvesvaraya National Institute of Technology (VNIT), Nagpur, India, for their guidance during research work; Prof. Jaroslav Stejskal, Head, Laboratory of Conducting Polymers, Institute of Macro Molecular Chemistry, Prague, Czech Republic; Prof. M.A. More, Department of Physics and Prof. A.A. Athawale, Department of Chemistry, University of Pune, India, Mr. Swapnil Deshpande, Dr. Sarin Kumar and Dr. Sudhakar Potukuchi, Eaton India Engineering Center, Pune for research collaboration; and Prof. V.S. Raja, Aqueous Corrosion Laboratory, Indian Institute of Technology, Mumbai, India, and Prof. R.D. Angal, renowned expert in the corrosion field, for their encouragement.

Finally, we would like to thank our families for their support, love, and understanding during the sometimes tedious process of preparing this book; Pravin P. Deshpande especially thanks his wife, Mrs. Swati P. Deshpande, for her patience and continuous encouragement. Moreover, the support provided by the team at CRC Press, particularly Allison Shatkin, Jill Jurgensen, and Adel Rosario, has been instrumental in publishing this book.

Pravin P. Deshpande
Dimitra Sazou

Authors

Pravin P. Deshpande obtained his PhD in metallurgical engineering from Visvesvaraya National Institute of Technology, Nagpur, and joined the Department of Metallurgy and Materials Science, College of Engineering Pune, in January 2008. Since then, he completed a number of research projects sponsored by University Grants Commission, New Delhi; Indian Space Research Organization, Bangalore; All India Council for Technical Education, New Delhi; Indian National Science Academy, New Delhi; and University of Pune and did consultancy work for a number of organizations. He has authored many refereed publications. He is an active member of a study group on Metallurgical Heritage of India, Indian National Academy of Engineering, New Delhi, India. His other interests include music and literature.

Dimitra Sazou is a professor of physical chemistry in the Department of Chemistry at the Aristotle University of Thessaloniki, Thessaloniki, Greece. She earned a PhD in chemistry from the Aristotle University of Thessaloniki. She was a visiting researcher/professor at the Department of Chemistry, University of Houston, and the Department of Physical Sciences, University of Cyprus. Prof. Sazou's research interests include corrosion–passivation of metals and their protection by using intrinsically conducting polymers, focusing on understanding the physico-electrochemical processes leading to uniform and localized corrosion via the analysis of the nonlinear behavior of metal electrodissolution–passivation reactions. She has more than 25 years of experience in the field and has authored many refereed publications.

List of Abbreviations

ABA	Aminobenzoic acid
ABF-G	Aminobenzoyl group-functionalized graphene
ABSA	Aminobenzenesulfonic acid
AC	Alternate current
ACAT	Aminocapped aniline trimer
AFM	Atomic force microscopy
AN	Aniline
AO	Aniline oligomer
AP	Adhesion promoter
APS	Ammonium persulfate
ATMP	Amino-trimethylene phosphonic
BS	Benzenesulfonate
CA	Carboxylic acid
CB	Conduction band
CE	Counter electrode
CNT	Single-walled carbon nanotubes
CP	Conducting polymers
CPE	Constant phase element
CPN	Conjugated polymer network
CPS	Cyclic potential sweep
CR	Corrosion rate
CrCC	Chromate conversion coatings
CRS	Cold-rolled steel
CS	Carbon steel
CSA	Camphorsulfonic acid
Cz	Carbazole
DBSA	Dodecylbenzenesulfonic acid
DC	Direct current
DoS	Dodecylsulfate
EB	Emeraldine base
E_g	Energy band gap
EIS	Electrochemical impedance spectroscopy
EMI	Electromagnetic interference
EPE	Epoxy ester
EPI	Electroactive polyimide
EQCM	Electrochemical quartz crystal microbalance
ES	Emeraldine salt
ESCA	Electron spectroscopy for chemical analysis
EWPU	Electroactive waterborne polyurethane
FE-SEM	Field emission scanning electron microscopy
FET	Field effect transistor
F-PAN	Fluoro-substituted polyaniline

FTIR	Fourier transform infrared spectroscopy
GC	Glassy carbon
HLB	Hydrophilic lipophilic balance
HPSC	Hydrophobic polyaniline-SiO_2 composite
HS	Hydrophobic surface
IR-RAS	Infrared reflection absorption spectroscopy
ITO	Indium tin oxide
LB	Leucoemeraldine base
LED	Light-emitting diode
LP	Lithium perchlorate
LPR	Linear polarization resistance
LS	Lingosulfonate
MEKFR	Methylethylketone formaldehyde resin
MMT	Montmorillonite
MS	Mild steel
MTMS	Methyl triethoxysilane
MWCNT	Multiwalled carbon nanotubes
NC	Nanocomposites
NEPI	Non-electroactive polyimide
NLO	Nonlinear optical
NMP	*N*-methyl-2-pyrrolidone
NP	Nanoparticle
NSA	*b*-naphthalenesulfonic acid
NSP	Nanospheres
OCP	Open circuit potential
ORR	Oxygen reduction reaction
P(3-AT)	Poly(3-alkylthiophene)
P(3-HT)	Poly(3-hexylthiophene)
P(3-MT)	Poly(3-methylthiophene)
P(3-OT)	Poly(3-octylthiophene)
PABS	Polyaminobenzene sulfonic acid
PAc	Polyacetylene
PACC	Polyaniline clay composite
PAGC	Polyaniline/graphene composite
PAN	Polyaniline
PANMA	Poly(aniline-*co*-metanilic acid)
PB	Pernigraniline base
PCC	Precipitated calcium carbonate
PCN	Polyimide-clay nanocomposite
PCz	Polycarbazole
PDA	Phenylenediamine
PDMS	Polydimethylsiloxane
PEDOT	Poly(3,4-ethylenedioxythiophene)
PFOS	Perfluorooctanesulfonate
PMo12	Phosphododecamolybdic acid
PNVCz	Poly(*N*-vinylcarbazole)

POT	Poly(*o*-toluidine)
PPy	Polypyrrole
PPV	Poly (para-phenylene vinylene)
PTC	PAN–TiO$_2$ composite
PTh	Polythiophene
PTSA	*p*-toluenesulfonic acid
PVA	Poly(vinyl alcohol)
PVAc	Poly(vinyl acetate)
PVC	Poly(vinyl chloride)
PVSS	Poly(vinylsulfonic acid) sodium salt
Py	Pyrrole
RE	Reference electrode
RoHS	Restriction of hazardous substance
SAM	Self-assembled monolayer
SCE	Saturated calomel electrode
SDS	Sodium dodecyl sulfate
SEM	Scanning electron microscopy
SHE	Standard hydrogen electrode
SH-PAN	Superhydrophobic polyaniline
SHS	Superhydrophobic surface
SP	Sodium perchlorate
SPAN	Sulfonated polyaniline
SS	Stainless steel
SVET	Scanning vibrating electrode technique
TBAP	Tetrabutylammonium perchlorate
TEAPFOS	Tetraethylammonium perfluorooctanesulfonate
TEM	Transmission electron microscopy
TGA	Thermogravimetric analysis
Th	Thiophene
VB	Valence band
WE	Working electrode
XPS	X-ray photoelectron spectroscopy
XRD	X-ray diffraction

1 An Overview of Recent Developments in Coating Anticorrosion Technology Based on Conducting Polymers

1.1 INTRODUCTION

In the last three decades, research on the anticorrosion properties of intrinsically conducting (or semiconducting) polymers, denoted throughout this book as conducting polymers (CPs), has flourished and led to the development of a wide variety of CP-based protective coatings with specifically designed properties. CPs are composed of conjugated chains containing π-electrons delocalized along the polymer backbone and combine physico-electrochemical properties that make them unique materials. They exhibit a wide range of conductivities (10^{-4} to 10^3 S cm^{-1}) in their p-doped (oxidized) state. The corrosion of metals is an electrochemical oxidative process during which metal releases electrons to an oxidizing species in the interface between metal and corrosive environment, resulting in the degradation or deterioration of the metal. The basic principle behind the idea to apply CPs for metal protection is related with this electron transfer process and passivity, by which nature has predicted the prevention or inhibition of metal degradation. DeBerry (1985) was the pioneer to indicate that polyaniline (PAN) was able to maintain the surface potential of PAN-coated stainless steel in sulfuric acid solutions into the passive state where a protective oxide film is formed on the alloy substrate. The key feature of this process is that PAN-based coatings are pinhole and defect tolerant in a similar way as the protective coatings based on the environmentally hazardous hexavalent chromium. Later on, Wessling (1996) showed how the oxidizing property of PAN is reestablished after metal oxidation by oxygen reduction within the CP layer.

Nowadays, CPs and CP-based composites/nanocomposites constitute a novel class of "smart" corrosion inhibiting coatings owing to their redox activity. Accumulation of a great number of studies on protective properties of CP-based coatings has provoked the appearance of several review articles. Early reviews provide surveys of work indicating the great interest on investigating CPs as potential inhibitive coatings for ferrous and nonferrous metals (McAndrew, 1997; Sitaram et al., 1997; Spinks et al., 2002; Tallman et al., 2002; Zarras et al., 2003a). Recent attempts summarize the advances made in the field during the last decade from different points of view

(Abu-Thabit & Makhlouf, 2014; Deshpande et al., 2014; Khan et al., 2010; Lacaze et al., 2010; Li & Wang, 2012; Ohtsuka, 2012; Rohwerder, 2009; Zarras & Stenger-Smith, 2014; Zarras et al., 2003b).

In this introductory chapter, an effort will be made to formulate an overview of the metal protection based on CPs by touching slightly on the breadth and depth of this field. Various aspects of the consequences of metallic corrosion, the necessity for metal protection, and the potential application of CPs in anticorrosion technology will be addressed more systematically in the following chapters. Section 1.2 refers to basic principles regarding the properties of CPs and especially their electronic conductivity from which a great variety of their applications stem. Emphasis is placed on the most widely used CPs in corrosion control. Section 1.3 deals with metal corrosion processes, to provide an idea of the problem arising from the destructive role of corrosion along with basic principles necessary for developing new materials that inhibit corrosion and extend the lifetime of metals. Section 1.4 outlines briefly the use of CPs to prevent or retard metal corrosion.

1.2   CONDUCTING POLYMERS

Since the breakthrough research regarding the doping of polyacetylene (PAc) published in 1977 (Shirakawa et al., 1977) by Alan J. Heeger, Alan G. MacDiarmid, and Hideki Shirakawa, interest in CPs has increased worldwide, leading to many innovative applications. The Nobel committee recognizing the seminal contribution of these mentioned scientists to the promising field of CPs awarded them jointly the Nobel Prize in Chemistry 2000 "for the discovery and development of conductive polymers" (Heeger, 2002; MacDiarmid, 2002; Shirakawa, 2002). It is interesting that, while at the end of the 20th century the interest in CPs had somehow settled down, this distinction to the field revitalized the interest of the scientific community in further investigating CPs.

Between 2000 and 2015, the number of publications on the topic indeed exhibited a tremendous increase. Because of the recognized potential for several important applications, PAN, polypyrrole (PPy), and polythiophene (PTh), along with their substituted derivatives, have a central role in the majority of the published studies (Chandrasekhar, 1999; Inzelt, 2008; Wallace et al., 2003). The chemical structure of selected CPs in their pristine (neutral) form is shown in Table 1.1. In their pristine form, CPs behave as insulators. Delocalized charge carriers accompanied by counter ions, so-called dopant ions, should be introduced in the π-conjugated polymer-chain to make CPs electronic conductors.

1.2.1   DOPING OF CPs

The term *doping* in the case of CPs refers to chemical oxidation/reduction processes and not in the sense that doping is utilized in the field of inorganic semiconductors. In the latter case, doping is called the addition of impurities into semiconductors. For example, adding a small amount of phosphorous with five valence electrons into semiconductor silicon with four valence electrons generates one excess electron for

TABLE 1.1

Chemical Structure and Conductivity, σ, of the Most Widely Utilized CPs in Metal Protection

CP	Chemical Structure (Neutral State)	σ (S cm^{-1}) (Doped State)
Polyaniline (PAN)		$10–10^3$
Polypyrrole (PPy)		$10^2–10^3$
Polythiophene (PTh)		10^2
Poly(para-phenylene vinylene) (PPV)		$10^3–10^4$
Polycarbazole (PCz)		10^{-4}

each phosphorous atom. At high temperatures, the excess electron is allowed to move in the conduction band (CB) of the silicon crystal. Such an impurity is called a donor and the semiconductor containing donors is defined as an n-type semiconductor. On the other hand, the addition of a small amount of boron with three valence electrons into semiconductor silicon generates one hole (vacant electron) for each boron atom, which, by thermal ionization, is allowed to move in the valence band (VB) of the silicon crystal. This kind of impurity is called acceptor and the semiconductor containing acceptors is defined as p-type semiconductor (Sato, 1998).

In the case of CPs, their oxidation involves the removal of electrons from the VB and the presence of charges on the CB of the polymer. The charges are delocalized over several polymer units, causing a relaxation of the geometry of the charged polymer to a more energetically favored conformation. This oxidation process resulting in the presence of positive charges and associated anions as counterions in the polymer chain is called p-type doping.

$$(CP)_n + nyMClO_4 \Leftrightarrow [(CP)^{+y}(ClO_4^-)_y]_n + nyM^+ + nye \qquad (1.1)$$

On the other hand, the reduction of CPs involves injection of electrons in the CB of the polymer, resulting in the presence of negative charges in the polymer chain and associated cations as counterions. This reduction process is called n-type doping.

$$(CP)_n + nyNaA + nye \Leftrightarrow [(CP)^{-y}(Na^+)_y]_n + nyA^- \qquad (1.2)$$

The M and A in reactions (1.1) and (1.2) denote the cation and anion, respectively. The inserted counterions, anions/cations during oxidation/reduction processes of CPs, are called dopants.

Dopants can be incorporated into the CP simultaneously with the synthesis. They may also be incorporated at a later stage or replaced by other desired ones. A reversible doping can be achieved via chemical, electrochemical, interfacial, and photo processes, as summarized in Figure 1.1, depending mostly on the application for which CPs have been designed to serve (Heeger, 2002). In the case of interfacial charge injection and photodoping, no dopant ions are involved. The type of doping process and the chemical nature of dopants constitute a tool for controlling the properties of CPs, such as redox and optical properties, morphology, processability, and environmental stability.

In the case of PAN, a special case of doping mechanism takes place based on acid-base chemistry. PAN exists in different oxidation states. Figure 1.2a shows the three base forms of PAN, namely, the colorless leucoemeraldine base (LB), a completely reduced form; the blue emeraldine base (EB), a partially oxidized state; and the violet pernigraniline base (PB), a fully oxidized state. All three PAN bases are insulating or semiconducting. Protonation by acids ($pK_a < 5.5$) results in the

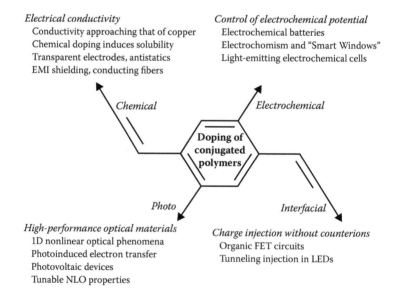

FIGURE 1.1 Doping mechanisms and related applications, where electromagnetic interference (EMI), nonlinear optical (NLO), field effect transistor (FET), and light-emitting diode (LED). (From Heeger, A.J., *Synth. Met.*, *125*, 23–42, 2002.)

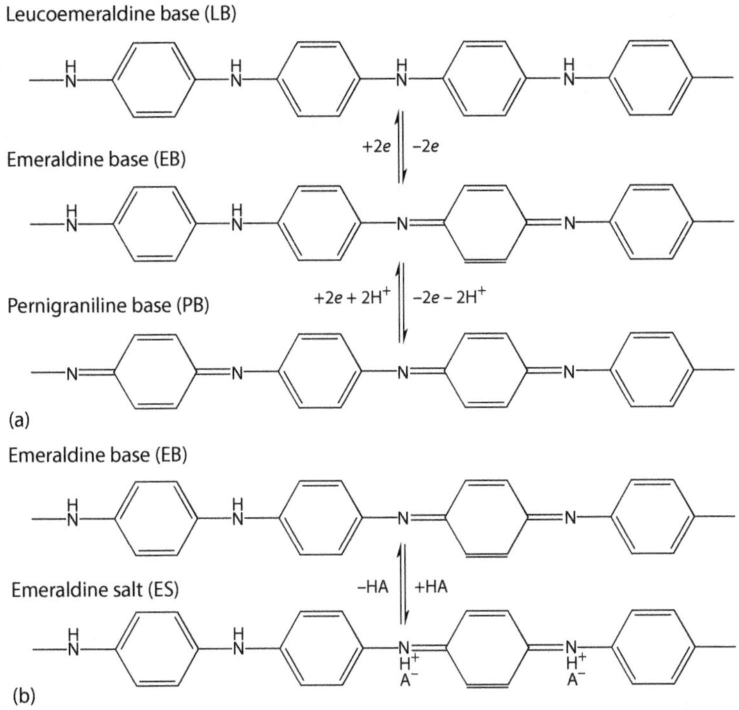

Leucoemeraldine base (LB)

Emeraldine base (EB)

$+2e \quad -2e$

Pernigraniline base (PB)

$+2e + 2H^+ \quad -2e - 2H^+$

(a)

Emeraldine base (EB)

Emeraldine salt (ES)

$-HA \quad +HA$

(b)

FIGURE 1.2 (a) Oxidation states (base forms) of PAN and (b) transformation of the nonconducting EB to the conducting ES form by proton acid doping.

corresponding salts of EB and PB. The protonation process (Figure 1.2b) by which the blue EB is transformed into the green emeraldine salt (ES) with a positive charge in each repeat unit associated with a counter anion triggers at the same time electronic conductivity in the PAN π-conjugated system, but without any change in the number of electrons. Therefore, among five oxidation states of PAN, only the ES form is conducting. Apart from protonation, chemical or electrochemical doping can be also applied to the LB to obtain the conducting ES form.

The electronic conductivity of CPs varies with increasing the extent of oxidation/ reduction in the polymer chain, the so-called doping level. The doping level is estimated as the proportion of dopant ions incorporated per monomer unit. Depending on the CP and the chemical nature of the dopant, the doping level cannot be higher than a certain value. It reaches a maximum value around 30%–40% for most of CPs. Maximum conductivity values can be seen in Table 1.1 for selected CPs.

A great variety of ions are used as dopants ranging from common small ions, such as Cl^-, ClO_4^-, HSO_4^-, and NO_3^- for p-type doping or Na^+ and Li^+ for n-type doping, to anions of a large size such as camphor-10-sulfonate (Silva et al., 2005, 2007), p-toluenesulfonate (Balaskas et al., 2011; Camalet et al., 1998), phosphonic acid (Kinlen et al., 2002), and even polyelectrolytes such as Nafion (Kosseoglou et al., 2011; Sazou & Kosseoglou, 2006; Sazou & Kourouzidou, 2009), poly(styrenesulfonate), or proteins and DNA (Skotheim & Reynolds, 2007).

1.2.2 Charge Transport in CPs

The mechanism of charge transport in CPs is described in terms of the energy band theory used in crystal materials, having in mind that the electronic conductivity in CPs is triggered by a different mechanism. As was previously mentioned, doping in CPs corresponds to oxidation/reduction processes. In the case of solid crystal materials, the energy state density of molecular orbital levels is very high resulting in the formation of wide orbital energy bands of bonding (lowest energy orbitals), nonbonding, and antibonding (highest energy orbitals) character. The energy bands are occupied by electrons successively from the lowest (inner orbitals) to the highest (frontier orbitals) level. Inner orbitals constitute localized bands attached to the lattice atoms whereas frontier bands are delocalized bands within the solid crystal. In conductors (metals), frontier energy bands are partially filled with electrons with the higher energy levels to be vacant, while electrons occupy successively the energy states from the lower band edge level to the Fermi level. In semiconductors, the frontier energy bands consist of the valence band (completely filled with electron low-energy bands) and conduction band (vacant of electron high-energy bands). Between the valence and conduction bands, there exists a forbidden gap, the energy band gap, E_g. For relatively narrow E_g, few electrons from the VB can move by thermal excitation to the CP, leaving holes (vacant electrons) in the VB. This process becomes impossible for a relatively wide E_g, as in the case of insulators. On the contrary, conductivity in metals is high because there is no band gap due to an overlap of the VB with the CB.

Doping in CPs may give rise to excitations like solitons, polarons, and bipolarons depending on the existence of degenerate or nondegenarate ground states of a CP (Bredas & Street, 1985; Moliton & Hiorns, 2004). When degenerate ground structure exists, as in trans-PAc, a neutral chain exhibits a defect (nonbonding orbital occupied by a single electron) as a result of isoenergetic regions of the structure. As seen in Figure 1.3, the left (L) and right (R) structures differ from each other only by the exchange of carbon–carbon

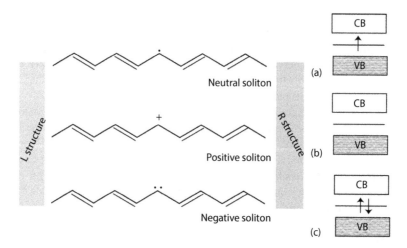

FIGURE 1.3 **(See color insert.)** (a) Neutral, (b) positive, and (c) negative solitons in trans-PAc along with their energy band structure indicating a half-occupied, a vacant, and a doubly occupied energy level (nonbonding) localized at the middle of the energy band gap, respectively. Arrows indicate electrons with spin direction, while charge-compensating anions are not shown.

single and double bonds. A neutral soliton has a 1/2 spin and zero charge. The spin density spreads over a number of carbons. This defect is a result of the so-called Peierls distortion (Peierls, 1955), which can be understood as follows: Starting from one side of the soliton, the double bonds become gradually longer and the single bonds shorter, in such a way that when arriving at the other side, the bond alternation is fully reversed. Considering an odd number of conjugated carbons, the single and double bonds become equal to each other at the middle of the soliton. A p-type doping of the neutral trans-PAc may result in cations (positive solitons) and insertion of charge-compensating anions, whereas an n-type doping results in anions (negative solitons) and insertion of charge-compensating cations along the polymer chain. The energy band structure of neutral, positive, and negative solitons is illustrated schematically in Figure 1.3.

Trans-PAc is unique in the degeneracy of its ground state because aromatic rings have no degenerate ground states. The energy of the neutral form with a benzenoid structure is lower as compared with the energy of the oxidized form with a quinoid structure. The removal of an electron from the p-conjugated system during the oxidation of the polymer creates a free radical cation with a 1/2 spin, as shown in Figure 1.4 for PPy. The radical cation is followed by a bond rearrangement resulting

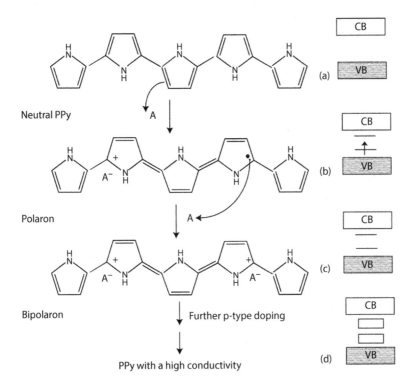

FIGURE 1.4 **(See color insert.)** (a) Neutral, (b) polaronic, and (c) bipolaronic states of PPy, with the corresponding energy band structures indicating generation of new localized electronic states within the energy band gap upon a p-type doping that results in the formation of positive polaron and bipolaron. (d) A further p-type doping of PPy leads to a higher conductivity, which results in the generation of new bands within the VB and CB.

in the formation of a quinoid-like bond sequence. The domain of the polymer chain with the quinoid structure is, however, limited because of its higher energy in comparison with the benzenoid structure dominating the remaining portion of the chain. For example, in the case of PPy, the lattice distortion is considered to extend over about four monomer units, and in PAN, over two monomer units. This coupling of a charged site with the free radical through a local distortion as a result polarization of the surrounding medium is called a polaron (Bredas & Street, 1985).

The neutral polymer has a full VB and a vacant CB, separated by a band gap (Figure 1.4a). The formation of a polaron corresponds to the creation of a new localized electronic state within the band gap, as shown in Figure 1.4b for the polaronic state of PPy. The polaron may be positive (radical cation), as in the case of p-type doping, or negative (radical anion), as in the case of n-type doping. A further doping of the CP leads to the formation of either bipolaron or two polarons. In the case of p-type doping of the polaronic state of a π-conjugated system, the removal of an electron may occur from either the polaron site, resulting in a dication (bipolaron), or somewhere in the rest of the polymer chain, resulting in the formation of two polarons. The formation of a bipolaron (Figure 1.4c) is coupled via a lattice distortion and a decrease in ionization energy and thus is energetically more favorable than the formation of two polarons. Quantum calculations for PPy predict a difference equal to 0.4 eV (Bredas & Street, 1985). A still further p-type doping generates bipolaron energy bands in the band gap. During continuous doping, the E_g increases, but because of merging of bipolaron bands within the VB and CP (Figure 1.4d), the polymer may reach even metal-like conductivities.

Both theoretical calculations and experimental studies support, to a sufficient extent, the previous rationalization of charge transport in CPs. It becomes clear that the nature of charge carriers (solitons, polarons, bipolarons) depends on the chemical nature of polymer and the optimum level of doping. Thereby, these factors, along with the doping method, type of dopant, and several environmental parameters (i.e., solvent, oxidant or applied potential, pH, temperature) associated with the CP synthesis, influence the electronic conductivity of CP. It is suggested that macroscopic charge transport in CPs represents a superposition of local transport mechanisms (1) within a conjugated chain, (2) from chain to chain, and (3) from fiber to fiber (Lyons, 1994; Roth & Bleier, 1987). Conduction mechanism 1 is referred for an intrinsic conductivity, while mechanisms 2 and 3, for a nonintrinsic conductivity. Intrinsic conductivity might be improved by ensuring that the CP has a sufficiently high molecular weight with well-aligned chains, containing minimum defects. Nonintrinsic conductivity (interchain and interfiber charge transport) might be described by hopping and tunneling models (Mott, 1987). The temperature dependence of the direct current (DC) and alternate current (AC) conductivity of CPs over a suitable temperature range may contribute essentially to the understanding of the valid conduction mechanism (Epstein, 1986; Kaiser, 2001).

1.2.3 APPLICATIONS OF CPs

The revived interest during the last 15 years has indeed led to the development of numerous innovative applications of CPs. A great impetus to this new technological direction was given by nanotechnology and a recent interest devoted to nanostructured

CPs, CP-based hybrid materials, and nanocomposites (Baibarac & Gomez-Romero, 2006; Bhandari et al., 2012; Ćirić-Marjanović, 2013; Gangopadhyay & De, 2000). Very promising CP-based nanostructured materials with improved properties, functionality, and commercial viability for various applications have been produced via several ways and processes (Eftekhari, 2010; Jang, 2006).

The electronic conductivity of CPs in conjunction with the processing properties of polymeric materials has drawn the interest of scientists and engineers from different disciplines, not necessarily associated with conventional polymers. Inspecting Figure 1.5, which illustrates evidently examples of exploited important applications of CPs, one recognizes the interdisciplinary nature of the field of CPs. PAN, PPy, and PTh dominate most technological applications. This is related primarily to the good stability of these CPs in a variety of conditions and their ease preparation by oxidative electrochemical or chemical polymerization in different oxidized states through adjusting the degree of p-type doping by modulating insertion/expulsion of dopants (Chujo, 2010).

All examples of applications included in Figure 1.5 are current areas of research. The optical and electronic properties of CPs were well exploited in electrochromic devices, light emitting diodes, and various electromagnetic interference shielding cases (Das & Prusty, 2012; Ponce de Leon et al., 2008; Skotheim & Reynolds, 2007; Wallace et al., 2003). Electronic along with reversible redox properties of CPs also offered a variety of applications in energy storage systems (Chen & Cheng, 2009; Dutta & Kundu, 2014; Whittingham, 2004) and electrocatalysis (Liu et al., 2008).

Another wide range of applications such as sensors, biosensors, and drug delivery systems is also based on the unique redox properties of CPs, which allow controlled ionic transport via the polymeric membrane triggered by various environmental parameters. Electrochemical switching of CPs between redox states is accompanied by the transport of dopant ions, in and out of the membrane for charge compensation. On the basis of these properties, a variety of anions have been electrostatically encapsulated into the CP membrane and released during reduction. On the other hand, by utilizing large anionic dopants that cannot move out of the membrane, cations have been released during oxidation of the CP. Therefore, the ability of CPs to

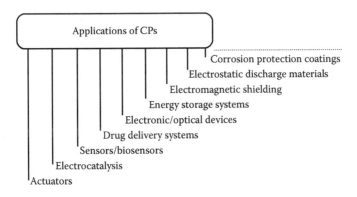

FIGURE 1.5 Examples of applications of CPs.

sense a chemical species led to several types of sensors (Ramanavicius et al., 2006; Vidal et al., 2003). In conjunction with this function, CPs can be suitably adjusted in order after chemical sensing to release another chemical species as a function of the concentration of the sensed species controlled via an external action (electrochemical triggered release), which led to drug delivery systems based on CPs (Geetha et al., 2006; Pernaut & Reynolds, 2000; Svirskis et al., 2010). Swelling/deswelling processes that accompanied the insertion/expulsion of dopant ions introduce also the challenge of mimicking the locomotion of living organisms and studying CPs as novel actuators (Otero, 2008; Pandey et al., 2003; Qi et al., 2004; Roemer et al., 2002). Moreover, CPs have been used in membrane-based separations owing to their inherent dynamical structure and controllable properties adjusted according to the redox state of the polymer (Martin et al., 1993; Sata et al., 1999; Zhao et al., 1998).

The intelligent release of dopant ions triggered by corrosion initiation on metal substrate and accompanied metal/CP interfacial changes in potential and pH is also based one of the most promising corrosion mechanisms suggested to rationalize the function of corrosion-resistant CP-based coatings. If corrosion inhibitors are chosen as dopants, the release from CP might inhibit corrosion or retard delamination originating from a defect (Barisci et al., 1998; Kendig et al., 2003; Kinlen et al., 1999, 2002; Paliwoda-Porebska et al., 2006).

However, it should be noted that in early studies on corrosion protection of metals by CPs, the semiconducting property seems to be the property that mostly attracted the interest of scientists and engineers. The idea behind this specific application was the exploitation of the semiconducting properties of CPs (in an analogy with the protective passive oxide film formed by metals in nature) along with their ability to provide "active" protection as the pigments containing the toxic hexavalent chromium. Toward the development of this idea, the fact that most CPs can be synthesized electrochemically by the anodic oxidation of suitable monomers such as pyrrole (Py), thiophene (Th), or aniline (AN), resulting in the formation of CP semiconducting films on a metal surface via a one-step process, helped to a great extent. Additionally, the fact that CP conductivity was readily controlled via electrochemical means reasonably led many electrochemists into the field. The field has gradually emerged since the first reports published by Mengoli et al. (1981), DeBerry (1985), and Ahmad and MacDiarmid (1996), exhibiting an accelerated progress since 1996, exemplified by the number of articles that appeared in literature and an interest of corrosion engineers to implement CPs in metal protection technology.

1.3   CORROSION OF METALS AND ITS PREVENTION

Corrosion is an electrochemical process occurring between metals and an aqueous corrosive environment with a catastrophic result, which is estimated to cost hundreds of billions of dollars annually. Although such estimates depend on time and place, values for the United States and United Kingdom are around 1%–3% of the gross national product at these countries, based on conservative estimates. Besides the huge economic consequences, corrosion affects many aspects of our societies as it deals with structural materials and infrastructure issues (bridges, pipelines, automobiles, airplanes, ships, etc.) and, hence, with safety and life-threatening situations. Although corrosion problems and its cost are, to some extent, inevitable because corrosion is the natural tendency of metals to revert to a more stable state, its consequences and cost can be remarkably reduced by proper design, choice of materials, and prevention methods. For a deeper understanding of corrosion processes, one needs to investigate both the electrochemical thermodynamics and kinetics of partial localized reactions, which are dealt with in Chapter 2.

There exist a large variety of corrosion forms that appear under different conditions depending mostly on the potential of the metal/corrosive environment interface as well as on the pH and the composition of the corrosive environment (Fontana, 2005). Examples of corrosion types include (1) generalized or uniform corrosion, (2) localized corrosion (pitting, crevice, filiform), (3) galvanic corrosion, (4) stress corrosion cracking, (5) intergranular corrosion, (6) selective leaching, (7) hydrogen damage (blistering, embrittlement), and (8) erosion-corrosion. The damage caused by several forms of corrosion varies in severity and can be classified into two general kinds, namely, aesthetic and engineering. Nevertheless, prevention or control of any kind of corrosion damage remains the ultimate goal, even in the stage of engineers' designs, wherever metals and alloys are used. It is of remarkable technological and economic importance. As is mentioned by Bockris and Reddy (2000), "It would be difficult to find a topic in technology in which the wide application of (even) present knowledge, would yield such a great financial gain." It is thus reasonable that after almost three centuries since the Faraday experiments, there still exists a constant interest from both scientific and practical aspects of corrosion and its prevention. Intensive research to gain a deeper understanding of the physico-electrochemical processes leading to metal corrosion and its prevention is still in progress.

It is generally accepted that the most important methods for preventing or retarding corrosion is, first, the choice of the suitable metal or alloy for a particular corrosive environment and, second, the proper design. Metal alloying has led to a variety of materials that are more resistant to corrosives and can be properly used at a relatively low cost considering their longer service life. A representative example is stainless steels, which represent different alloys of iron containing mainly chromium and nickel at various proportions (Sedriks, 1996). However, despite stainless steel improving corrosion resistance, under certain aggressive corrosive conditions (i.e., in the presence of chlorides), stainless steels are susceptible to various types of localized corrosion, with catastrophic results. Although several optimum combinations of metal/corrosive systems presenting the maximum corrosion resistance for the least economical cost are suggested, their utilization is not always feasible. Neither is the addition of corrosion inhibitors (organic or inorganic substances) feasible in all cases, which represents another important way to prevent or retard the spontaneous electrochemical dissolution of metals and alloys. Besides proper selection and design in metal use, the main methods utilized for corrosion prevention are classified in three main classes:

- Corrosion inhibitors. An easy way is the addition of suitable substances (inhibitors) in the solution in contact with the metal. However, since this is not a feasible way in cases where metal structures are exposed in atmospheric humid conditions and/or in marine environments, other ways were devised in applying corrosion inhibitors. Corrosion inhibitors are encapsulated in coatings and are released from the coatings by a leaching mechanism. A subsequent reaction between the inhibitor and the substrate metal results in the formation of a protective layer that prolongs corrosion protection.
- Electrochemical control. This is achieved by passing cathodic or anodic current into the metal. Cathodic protection is achieved by supplying electrons to the metal to suppress the anodic partial reaction, namely, metal dissolution, and reinforce the rate of cathodic partial reaction (hydrogen evolution). This can be implemented by an external power supply or by suitable galvanic coupling using sacrificial anodes such as zinc, aluminum, or magnesium. Anodic protection is realized by the formation of a protective layer on metal surfaces when an appropriate anodic current is imposed externally. It is efficient mainly in metals that exhibit an active-to-passive transition such as Fe, Ni, Cr, and their alloys.
- Organic/inorganic coatings. These aim to provide isolation of the metal surface from the corrosive environment by organic/inorganic coatings acting as physical barriers and inhibit corrosion (Munger & Vincent, 1999). Coatings could be classified into two main groups, active coatings such as sacrificial anodes, which function even when parts of the substrate metal are exposed to the corrosive environment, and barrier coatings, such as paints that prevent corrosion by isolating the metal from the corrosive environment. Paints of various types embrace a simple and effective method for corrosion prevention as long as the barrier remains intact. Once they are damaged, galvanic corrosion may begin, as in the case of coatings by noble metals, or permeation of oxygen and H_2O may be allowed, as in the case of nonmetallic coatings.

A variety of formulations of organic coatings and paints are extensively applied for an efficient protection of a wide range of metals against corrosion (Wicks et al., 2007). A coating with an effective protection performance is a "complex" system consisting of successive layers of a different thickness designed to serve a specific role in protecting metals (Marrion, 2004). In practice, a primer with inhibiting properties of about 1 μm in thickness is applied first to ensure adhesion to the metal. A relatively thick polymer layer of 20–30 μm in thickness, which contains pigments and anticorrosion agents, follows the primer. Both layers are cured to achieve good cross-linking between polymer chains while a thin topcoat layer of an appropriate composition is applied for aesthetic purposes. The selection of major components and their combination with other substances added in low quantities, like surface-active agents or biocides, influences the overall performance of the coating. Coating technology exhibits a rapid development during recent decades focusing in self-healing materials (Fedrizzi, 2011) and environmentally friendly inhibitors, so-called "green inhibitors" (Hughes et al., 2010).

Although the term *self-healing* was not used in protective coatings until the 1990s, a traditional way of using corrosion inhibitors that are released upon leaching as described previously can be characterized as a self-healing mechanism. Additionally, chromate-based coatings with excellent corrosion properties for several metals exhibit also a self-healing property based on the different oxidation states of chromium. Chromate is an oxidizing oxyanion of hexavalent chromium, Cr(VI), that is toxic and carcinogenic. Because Cr(VI) is water soluble, it can migrate and accumulate within the environment.

Cr(VI)-based anticorrosive technologies are superior in protection effectiveness and are hence still in use in manufacturing worldwide. The Occupational Safety and Health Administration in the United States states that Cr(VI) exposure results in an increased risk of lung cancer, and in addition, occupational exposure may lead to asthma and damage to the nasal lining. The European Union adopted the Restriction of Hazardous Substances directive (RoHS) in February 2003, which includes Cr(VI) as one of the hazardous materials. The RoHS directive took effect on July 1, 2006, and needs to be enforced and become law in each European Union member country. Current regulations regarding industries that use corrosion control technologies are intended to prevent environmental contamination and protect human health. It is not certain if these regulations and practices do, in fact, achieve a satisfactory level of protection. Another environmental motivation for change in the coatings industry is reduction of the use of volatile organic compounds such as solvents.

Recognizing that hexavalent chromium regulations are going to be enhanced, alternative nontoxic inexpensive materials with self-healing properties have been suggested. Thus, new anticorrosive technologies have arisen during the last decades using smart materials that can replace the Cr(VI)-based ones. CPs such as PAN, PPy, PTh, and their derivatives are among the key components in various novel protective-coating systems. Figure 1.6 illustrates a typical chromate-based multilayered protective coating system used in aerospace in comparison with an alternative chromate-free coating system with similar or even improved functionalities (Hughes et al., 2010). A CP-based protective coating system will combine the function of barrier and self-healing and may contain (e.g., as dopants) additional self-healing agents. It may respond to environmental changes by changes in structure and properties, often reversibly. Stimulus and response sensing mechanisms are still obscure and further understanding is required.

Nanostructured materials engineering opened the possibility of engineering a new class of smart coatings with green inhibitors and self-healing properties. The ultimate goal of recent research is the design of nanostructured smart coatings engineered to provide a superior resistance to corrosion on demand, when the coating is breached or when an electrical or mechanical control signal is applied to the coating, while at the same time other functionalities like barrier protection or aesthetic properties could be recovered. Within this context, CPs in coatings technology appears to be a promising approach.

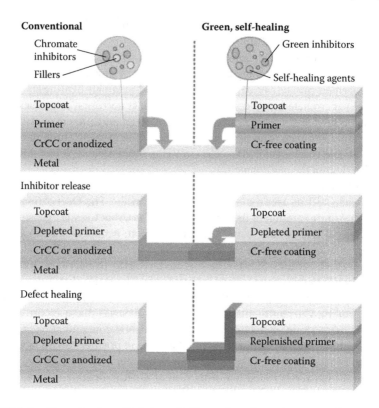

FIGURE 1.6   (See color insert.) Schematic illustration of postdamage healing in conventional chromate conversion coatings (CrCC) (left) and smart, green (right) self-repairing coatings. The structure of a smart, green coating might comprise CPs and green inhibitors as dopants that would release into a defect upon exposure to a corrosive environment. (From Hughes, A.E., Cole, I.S., Muster, T.H., Varley, R.J., *NPG Asia Mater.*, 2, 143–151, 2010.)

1.4   CORROSION PROTECTION BY CPs

CPs have been considered as key components for a new generation of active coatings with self-healing properties for metal corrosion control. The concept is based primarily on CP's semiconducting properties combined with processing advantages of conventional polymers. It is worth mentioning that semiconductive oxide layers formed on metal surfaces is the way by which nature has predicted the protection and long life of metals exposed in moisture environments. First, a distinction should be made between intrinsically conductive polymers and electroactive polymers. Electroactive polymers comprise a wide range of materials, which include redox polymers, loaded ionomers, and CPs. This book mostly addresses the utilization of CPs in metal anticorrosion control, capitalizing their intrinsic conductivity due to electron delocalization along the polymer backbone through a π-conjugated system.

The most widely used CPs in metal corrosion protection can be classified under PAN, PPy, PTh, poly(para-phenylene vinylene) (PPV), and polycarbazole (PCz), the chemical structure of which, along with a range of their conductivities, is summarized in Table 1.1. Their synthesis can be carried out relatively easily through a variety of chemical and electrochemical methods. To improve the properties of CPs to make them suitable for anticorrosion coatings, a variety of derivatives of AN, Py, and Th were synthesized by a proper choice of substituents and synthesis conditions. For a CP-based coating to qualify as a candidate for corrosion protection, selection criteria include processability, stability, ease of preparation, adhesion to the metal-substrate, and long-term mechanical integrity.

Comparing between different classes of CPs used for the protection of metals against corrosion, PAN and PPy seem to be the most frequently studied ones. PAN is being used more because of its better environmental stability, lower cost, easier synthesis, and diversity in oxidation states owing to its protonic acid doping (Figure 1.2b). All other CPs exhibit three conducting states, namely, neutral (uncharged), with insulating or semiconducting properties; oxidized (p-doped), with conducting or semiconducting properties; and reduced (n-doped), with insulating or semiconducting properties. In the oxidized and reduced states, electrons are added or removed, respectively, from the polymer chain of the neutral polymer. As mentioned previously, CPs can be converted from one state to another via different doping methods (Figure 1.1). The characteristic ability of CPs to store and transport charge is the basis of their functionality, classified as "smart" protection materials. Within this respect, effective anticorrosion protection can be achieved when they are in the doped state, although there are reports indicating effectiveness also for CP's undoped state (Fahlman et al., 1997; McAndrew, 1997; Talo et al., 1999).

Despite the large number of studies carried out in CPs regarding their potential application in the protection of a variety of metals (ferrous and nonferrous) against corrosion when in contact with acid and chloride-containing media, our understanding of the mechanism by which CPs provide protection is quite limited. The protection mechanism seems to depend on various factors, including the type of CP, its formulation when acting as coating, and metal substrate. CPs applied in their oxidized state may act as oxidizers either by enhancing the oxide layer underlying the polymer or by healing the passive state in small defects. This mechanism was early on suggested and supported through a myriad of evidences, in particular, in the case of PAN protecting ferrous metals, which exhibit active to passive transition. Such an anodic protection mechanism requires that the reduced CP can be reoxidized by an oxidizing agent, i.e., by the atmospheric O_2, for the charge consumed by metal oxidation to be replenished within the CP layer. In this respect, PAN and PPy can be oxidized by O_2, while PTh is not. PAN and PPy are stable in air in their oxidized state, whereas PTh, in its neutral state. For example, electrodeposited polymethylthiophen films reduce the corrosion rate of mild steel, but they cannot passivate the metal substrate (Rammelt et al., 2001).

PAN can be utilized as either a corrosion inhibitor or a protective coating (Benchikh et al., 2009; Cook et al., 2004). It can function as an inhibitor because it is adsorbed on the metal surface through the functional group C = N. The situation is more complicated when it provides protection as coating, and many studies have been devoted to clarify the mechanism for the corrosion protection of PAN coatings. Most of the experimental findings agree that protection of iron and stainless steel can be realized by doped PAN, which is reduced by oxidizing the underlying metal surface (Bernard et al., 1999, 2001; Gasparac & Martin, 2001; Sazou, 2001; Sazou & Georgolios, 1997; Wessling, 1994). Contradictory results exist in the literature regarding which one is the more efficient form of PAN in metal protection and if PAN does protect in chloride-containing solutions.

However, PAN loses its electroactivity at pH>4 and, hence, its smart function, acting solely as a physical barrier. PAN and other CPs can be applied as protective coatings for corrosion protection either as primers or as components of a barrier organic coating in a variety of formulations (blends, composites/nanocomposites) or as additives to improve protective efficiency and long-term active function of conventional coatings. Bi/multilayers and copolymers of CPs have been also utilized within the context of improving the protective efficiency of CPs in certain cases of metals and alloys.

Overall, the mechanism of protection is more complex than that of conventional organic coatings acting mostly as barriers to increase the diffusion resistance of corrosive species toward the metal substrate. On the basis of the knowledge acquired through intensive research developed during the last decades, CP-based coatings can provide protection by several ways, as Figure 1.7 concisely shows: (1) as an active electronic barrier, (2) as a physical barrier, (3) as a self-healing agent, including anodic protection by "ennobling" the underlying metal and intelligent release of corrosion inhibiting anions stored in the redox polymer, (4) as cathodic protection, (5) displacement of oxygen reduction, and (6) as a mediator/catalyst for the formation of corrosion products that form inert passive layers.

Most of the known protective mechanisms were suggested on the basis of evaluation results regarding the anticorrosion properties of PAN-based coatings. Different mechanisms seem to proceed in parallel depending on the CP, metal substrate, and corrosive environment. The unique properties of CPs make them very suitable candidates for the development of multifunctional coatings, as the ultimate goal of automotive, aerospace, and other manufacturers is the design of self-healing coatings

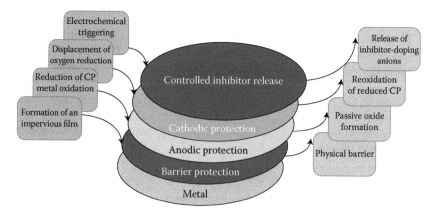

FIGURE 1.7   (See color insert.) Summary of possible mechanisms by which CPs may protect metals against corrosion.

that could be recovered after mechanical and chemical damages. CP-based coatings are considered as multifunctional coatings in the sense that they may offer more than one function to the metal, such as self-healing properties, corrosion indicators, physical barriers, or antistatic and antifouling properties as composites. Within this frame, the genesis of a new generation of CP-based coatings that can offer more than an anticorrosion protection to the metal is under progress, as seen by numerous publications exploring CP composites and nanocomposites. CP composites/nano-composites combine CP properties with characteristics of other inorganic or organic materials, resulting in innovative smart coatings of a better quality and a lower cost without undesired effects caused when CP primers are utilized alone. However, further efforts are required toward a better understanding of the potential and limitations of CP-based smart coatings in anticorrosion technology.

REFERENCES

Abu-Thabit, N. Y., & Makhlouf, A. S. H. (2014). Recent advances in polyaniline (PAN)-based organic coatings for corrosion protection. In: A. S. H. Makhlouf, ed., *Handbook of Smart Coatings for Materials Protection* (pp. 459–486). Cambridge, UK: Woodhead Publishing Ltd.

Ahmad, N., & MacDiarmid, A. G. (1996). Inhibition of corrosion of steels with the exploitation of conducting polymers. *Synth. Met.*, 78(2), 103–110.

Baibarac, M., & Gomez-Romero, P. (2006). Nanocomposites based on conducting polymers and carbon nanotubes from fancy materials to functional applications. *J. Nanosci. Nanotechnol.*, 6, 1–14.

Balaskas, A. C., Kartsonakis, I. A., Kordas, G., Cabral, A. M., & Morais, P. J. (2011). Influence of the doping agent on the corrosion protection properties of polypyrrole grown on aluminium alloy 2024-T3. *Prog. Org. Coat.*, 71(2), 181–187.

Barisci, J. N., Lewis, T. W., Spinks, G. M., Too, C. O., & Wallace, G. G. (1998). Conducting polymers as a basis for responsive materials systems. *J. Intell. Mater. Syst. Struct.*, 9, 723–731.

Benchikh, A., Aitout, R., Makhloufi, L., Benhaddad, L., & Saidani, B. (2009). Soluble conducting poly(aniline–co-orthotoluidine) copolymer as corrosion inhibitor for carbon steel in 3% NaCl solution. *Desalination*, 249, 466–474.

Bernard, M. C., Hugot-Le Goff, A., Joiret, S., Dinh, N. N., & Toan, N. N. (1999). Polyaniline layer for iron protection in sulfate medium. *J. Electrochem. Soc.*, 146(3), 995–998.

Bernard, M. C., Joiret, S., Hugot-Le Goff, A., & Long, P. D. (2001). Protection of iron against corrosion using a polyaniline layer II. Spectroscopic analysis of the layer grown in phosphoric/metanilic solution. *J. Electrochem. Soc.*, 148, B299–B303.

Bhandari, H., Anoop Kumar, S., & Dhawan, S. K., eds. (2012). *Conducting Polymer Nanocomposites for Anticorrosive and Antistatic Applications*. Rejeka, Croatia: INTECHOPEN.COM: INTECH.

Bockris, J. O. M., & Reddy, A. K. N. (2000). *Modern Electrochemistry 2B*. New York: Kluwer Academic/Plenum Publishers.

Bredas, J. L., & Street, G. B. (1985). Polarons, bipolarons and solitons in conducting polymers. *Acc. Chem. Res.*, 18, 309–315.

Camalet, J. L., Lacroix, J. C., Aeiyach, S., & Lacaze, P. C. (1998). Characterization of polyaniline films electrodeposited on mild steel in aqueous p-toluenesulfonic acid solution. *J. Electroanal. Chem.*, 445, 117–124.

Chandrasekhar, P. (1999). *Conducting Polymers, Fundamentals and Applications: A Practical Approach*. Norwell, MA: Kluwer Academic Publishers.

Chen, J., & Cheng, F. (2009). Combination of lightweight elements and nanostructured materials for batteries. *Acc. Chem. Res.*, *42*, 713–723.

Chujo, Y. (2010). *Conducting Polymer Synthesis—Methods and Reactions*. Weinheim: Wiley-VCH Verlag GmbH & Co. KGaA.

Ćirić-Marjanović, G. (2013). Recent advances in polyaniline composites with metals, metalloids and nonmetals. *Synth. Met.*, *170*, 31–56.

Cook, A., Gabriel, A., Siew, D., & Laycock, N. (2004). Corrosion protection of low carbon steel with polyaniline: Passivation or inhibition? *Curr. Appl. Phys.*, *4*, 133–136.

Das, T. K., & Prusty, S. (2012). Review on conducting polymers and their applications. *Polym. Plast. Technol. Eng.*, *51*, 1487–1500.

DeBerry, D. W. (1985). Modification of the electrochemical and corrosion behavior of stainless steels with an electroactive coating. *J. Electrochem. Soc.*, *132*(5), 1022–1026.

Deshpande, P. P., Jadhav, N. G., Gelling, V. J., & Sazou, D. (2014). Conducting polymers for corrosion protection: A review. *J. Coat. Technol. Res.*, *11*, 473–494.

Dutta, K., & Kundu, P. P. (2014). A review on aromatic conducting polymers-based catalyst supporting matrices for application in Microbial fuel cells. *Polym. Rev.*, *54*, 401–435.

Eftekhari, A., ed. (2010). *Nanostructured Conductive Polymers*. Chichester, UK: John Wiley & Sons Ltd.

Epstein, A. J. (1986). AC conductivity of polyacetylene distinguishing mechanisms of charge transport. In: T. Skotheim, & J. Reynolds, eds., *Handbook of Conducting Polymers*, vol. 2 (p. 1041). New York: Marcel Dekker.

Fahlman, M., Jasty, S., & Epstein, A. J. (1997). Corrosion protection of iron/steel by emeraldine base polyaniline: An X-ray photoelectron spectroscopy study. *Synth. Met.*, *85*, 1323–1326.

Fedrizzi, L. (2011). *European Federation of Corrosion Series, vol. 58: Self-Healing Properties of New Surface Treatments*. Wakefield, UK: Maney Publishing.

Fontana, M. G. (2005). *Corrosion Engineering*. New Delhi: Tata McGraw Hill Education Private Ltd.

Gangopadhyay, R., & De, A. (2000). Conducting polymer nanocomposites: A brief overview. *Chem. Mater.*, *12*, 608–622.

Gasparac, R., & Martin, C. R. (2001). Investigations of the mechanism of corrosion inhibition by polyaniline—Polyaniline-coated stainless steel in sulfuric acid solution. *J. Electrochem. Soc.*, *148*(4), B138–B145.

Geetha, S., Rao, C. R. K., Vijayan, M., & Trivedi, D. C. (2006). Biosensing and drug delivery by polypyrrole. *Anal. Chim. Acta*, *568*(1–2), 119–125.

Heeger, A. J. (2002). Semiconducting and metallic polymers: The fourth generation of polymeric materials. *Synth. Met.*, *125*, 23–42.

Hughes, A. E., Cole, I. S., Muster, T. H., & Varley, R. J. (2010). Designing green, self-healing coatings for metal protection. *NPG Asia Mater.*, *2*, 143–151.

Inzelt, G. (2008). *Conducting Polymers: A New Era in Electrochemistry*. Berlin, Heidelberg: Springer.

Jang, J. (2006). Conducting polymer nanomaterials and their applications. *Adv. Polym. Sci.*, *199*, 189–259.

Kaiser, A. B. (2001). Systematic conductivity behavior in conducting polymers: Effects of heterogeneous disorder. *Adv. Mater.*, *13*, 927–941.

Kendig, M., Hon, M., & Warren, L. (2003). "Smart" corrosion inhibiting coatings. *Prog. Org. Coat.*, *47*, 183–189.

Khan, M. I., Chaudhry, A. U., Hashim, S., Zahoor, M. K., & Igbal, M. Z. (2010). Recent developments in intrinsically conductive polymer coatings for corrosion protection. *Chem. Eng. Res. Bull.*, *14*, 73–86.

Kinlen, P. J., Menon, V., & Ding, Y. W. (1999). A mechanistic investigation of polyaniline corrosion protection using the scanning reference electrode technique. *J. Electrochem. Soc.*, *146*(10), 3690–3695.

Kinlen, P. J., Ding, Y., & Silverman, D. C. (2002). Corrosion protection of mild steel using sulfonic and phosphonic acid-doped polyanilines. *Corrosion, 58*(6), 490–497.

Kosseoglou, D., Kokkinofta, R., & Sazou, D. (2011). FTIR spectroscopic characterization of NafionA (R)-polyaniline composite films employed for the corrosion control of stainless steel. *J. Solid State Electrochem., 15*(11–12), 2619–2631.

Lacaze, P. C., Ghilane, J., Randriamahazaka, H., & Lacroix, J.-C. (2010). Electroactive conducting polymers for the protection of metals against corrosion: From micro- to nanostructured films. In: A. Eftekhari, ed., *Nanostructured Conductive Polymers*. Chichester, UK: John Wiley & Sons, Ltd.

Li, Y., & Wang, X. (2012). Intrinsically conducting polymers and their composites for anticorrosion and antistatic applications. In: X. Yang, ed., *Semiconducting Polymer Composites: Principles, Morphologies, Properties and Applications* (pp. 269–298). Weinheim, Germany: Wilwy-VCH Verlag GmbH & Co. KGaA.

Liu, F.-J., Huang, L.-M., Wen, T.-C., Li, C.-F., Huang, S.-L., & Gopalan, A. (2008). Effect of deposition sequence of platinum and ruthenium particles into nanofibrous network of polyanilinepoly(styrene sulfonic acid) on electrocatalytic oxidation of methanol. *Synth. Met., 158*, 603–609.

Lyons, M. E. G. (1994). Charge percolation in electroactive polymers. In: M. E. G. Lyons, ed., *Electroactive Polymer Chemistry. Part 1 Fundamentals*, vol. 1 (pp. 1–226). New York: Plenum Press.

MacDiarmid, A. G. (2002). Synthetic metals: A novel role of organic polymers. *Synth. Met., 125*, 11–22.

Marrion, A. R. (2004). *The Chemistry and Physics of Coatings*. Cambridge, UK: Royal Society of Chemistry.

Martin, C. R., Liang, W., Menon, V., Parthasarathy, R., & Parthasarathy, A. (1993). Electronically conductive polymers as chemically-selective layers for membrane-based separations. *Synth. Met., 57*, 3766–3773.

McAndrew, T. P. (1997). Corrosion prevention with electrically conductive polymers. *TRIP, 5*(1), 7–12.

Mengoli, G., Munari, M. T., Bianco, P., & Musiani, M. M. (1981). Anodic synthesis of polyaniline coatings onto Fe surfaces. *J. Appl. Polym. Sci., 26*, 4247–4257.

Moliton, A., & Hiorns, R. C. (2004). Review of electronic and optical properties of semiconducting π-conjugated polymers: Applications in optoelectronics. *Polymer Int., 53*, 1397–1412.

Mott, N. F. (1987). *Conduction in Noncrystalline Materials*. Oxford: Oxford University Press.

Munger, C. G., & Vincent, L. D. (1999). *Corrosion Prevention by Protective Coatings*. Houston, TX: NACE International.

Ohtsuka, T. (2012). Corrosion protection of steels by conducting polymers. *Int. J. Corros., 2012*, 1–7.

Otero, T. F. (2008). Artificial muscles, sensing and multifunctionality. In: M. Shahinpoor, & H.-J. Schenider, eds., *Intelligent Materials* (pp. 142–190). Cambridge, UK: Royal Society of Chemistry.

Paliwoda-Porebska, G., Rohwerder, M., Stratmann, M., Rammelt, U., Duc, L. M., & Plieth, W. (2006). Release mechanism of electrodeposited polypyrrole doped with corrosion inhibitor anions. *J. Solid State Electrochem., 10*, 730–736.

Pandey, S. S., Takashima, W., & Kaneto, K. (2003). Structure property correlation: Electrochemomechanical deformation in polypyrrole films. *Thin Solid Films, 438*, 206–211.

Peierls, R. E. (1955). *Quantum Theory of Solids*. London: Oxford University Press.

Pernaut, J.-M., & Reynolds, J. R. (2000). Use of conducting electroactive polymers for drug delivery and sensing of bioactive molecules. A redox chemistry approach. *J. Phys. Chem. B, 104*, 4080–4090.

Ponce de Leon, C., Campbell, S. A., Smith, J. R., & Walsh, F. C. (2008). Conducting polymer coatings in electrochemical technology Part 2—Application areas. *Trans. Inst. Met. Finish.*, *86*, 34–40.

Qi, B. H., Wu, L., & Mattes, B. R. (2004). Strain and energy efficiency of polyaniline fiber electrochemical actuators in aqueous electrolytes. *J. Phys. Chem. B*, *108*, 6222–6227.

Ramanavicius, A., Ramanaviciene, A., & Malinauskas, A. (2006). Electrochemical sensors based on conducting polymer—Polypyrrole. *Electrochim. Acta*, *51*(27), 6025–6037.

Rammelt, U., Nguyen, P. T., & Plieth, W. (2001). Protection of mild steel by modification with thin films of polymethylthiophene. *Electrochim. Acta*, *46*(26–27), 4251–4257.

Roemer, M., Kurzenknabe, T., Oesterschulze, E., & Nicoloso, N. (2002). Microactuators based on conducting polymers. *Anal. Bioanal. Chem.*, *373*, 754–757.

Rohwerder, M. (2009). Conducting polymers for corrosion protection: A review. *Int. J. Mater. Res.*, *100*(10), 1331–1342.

Roth, S., & Bleier, H. (1987). Solitons in polyacetylene. *Adv. Phys.*, *36*, 385.

Sata, T., Ishii, Y., Kawamura, K., & Matsusaki, K. (1999). Composite membranes prepared from cation-exchange membranes and polyaniline and their transport properties in electrodialysis. *J. Elecrtrochem. Soc.*, *146*, 585–591.

Sato, N. (1998). *Electrochemistry of Metal and Semiconductor Electrodes*. Amsterdam, The Netherlands: Elsevier Science B.V.

Sazou, D. (2001). Electrodeposition of ring-substituted polyanilines on Fe surfaces from aqueous oxalic acid solutions and corrosion protection of Fe. *Synth. Met.*, *118*, 133–147.

Sazou, D., & Georgolios, C. (1997). Formation of conducting polyaniline coatings on iron surfaces by electropolymerization of aniline in aqueous solutions. *J. Electroanal. Chem.*, *429*(1–2), 81–93.

Sazou, D., & Kosseoglou, D. (2006). Corrosion inhibition by Nafion (R)-polyaniline composite films deposited on stainless steel in a two-step process. *Electrochim. Acta*, *51*(12), 2503–2511.

Sazou, D., & Kourouzidou, M. (2009). Electrochemical synthesis and anticorrosive properties of Nafion®–poly(aniline-co-o-aminophenol) coatings on stainless steel. *Electrochim. Acta*, *54*(9), 2425–2433.

Sedriks, A. J. (1996). *Corrosion of Stainless Steels*. New York: John Wiley & Sons, Inc.

Shirakawa, H. (2002). The discovery of polyacetylene film. The dawning of an era of conducting polymers. *Synth. Met.*, *125*, 3–10.

Shirakawa, H., Louis, E., MacDiarmid, A. G., & Heeger, A. J. (1977). Synthesis of electrically conducting organic polymers: Halogen derivatives of polyacetylene, (CH)x. *Chem. Commun.*, 578–588.

Silva, J. E. P., Torresi, S. I. C., & Torresi, R. M. (2005). Polyaniline acrylic coatings for corrosion inhibition: The role played by counter-ions. *Corros. Sci.*, *47*, 811–822.

Silva, J. E. P., Torresi, S. I. C., & Torresi, R. M. (2007). Polyaniline/poly(methylmethacrylate) blends for corrosion protection: The effect of passivating dopants on different metals. *Prog. Org. Coat.*, *58*, 33–39.

Sitaram, S. P., Stoffer, J. O., & O'Keefe, T. J. (1997). Application of conducting polymers in corrosion protection. *J. Coat. Technol.*, *69*, 65–69.

Skotheim, T., & Reynolds, J. (2007). *Handbook of Conducting Polymers*. New York: CRC Press.

Spinks, G. M., Dominis, A. J., Wallace, G. G., & Tallman, D. E. (2002). Electroactive conducting polymers for corrosion control—Part 2. Ferrous metals. *J. Solid State Electrochem.*, *6*(2), 85–100.

Svirskis, D., Travas-Sejdic, J., Rodgers, A., & Garg, S. (2010). Electrochemically controlled drug delivery based on intrinsically conducting polymers. *J. Control. Release*, *146*(1), 6–15.

Tallman, D. E., Spinks, G., Dominis, A., & Wallace, G. G. (2002). Electroactive conducting polymers for corrosion control Part 1. General introduction and a review of non-ferrous metals. *J. Solid State Electrochem.*, *6*(2), 73–84.

Talo, A., Forsen, O., & Ylasaar, S. (1999). Corrosion protective polyaniline epoxy blend coating on mild steel. *Synth. Met.*, *102*, 1394–1395.

Vidal, J.-C., Garcia-Ruiz, E., & Castillo, J.-R. (2003). Recent advances in electropolymerized conducting polymers in amperometric biosensors. *Microchim. Acta*, *143*, 93–111.

Wallace, G. G., Spinks, G. M., Kane-Maguire, L. A. P., & Teasdale, P. R. (2003). *Conductive Electroactive Polymers-Intelligent Materials Systems*. New York: CRC Press.

Wessling, B. (1994). Passivation of metals by coating with polyaniline—Corrosion potential shift and morphological-changes. *Adv. Mater.*, *6*(3), 226–228.

Wessling, B. (1996). Corrosion prevention with an organic metal (polyaniline): Surface ennobling, passivation, corrosion test results. *Werkst. Korros.-Mater. Corros.*, *47*(8), 439–445.

Whittingham, M. S. (2004). Lithium batteries and cathode materials. *Chem. Rev.*, *104*, 4271–4301.

Wicks, Z. W. J., Jones, F. N., Pappas, S. P., & Wicks, D. A. (2007). *Organic Coatings, Science and Technology*. Hoboken, NJ: John Wiley & Sons, Inc.

Zarras, P., Anderson, N., Webber, C., Irvin, D. J., Irvin, J. A., Guenthner, A., & Stenger-Smith, J. D. (2003a). Progress in using conductive polymers as corrosion-inhibiting coatings. *Rad. Phys. Chem.*, *68*, 387–394.

Zarras, P., Stenger-Smith, J. D., & Wei, Y. (2003b). *Electroactive Polymers for Corrosion Control*. Washington, DC: American Chemical Society.

Zarras, P., & Stenger-Smith, J. D. (2014). Electro-active polymer (EAP) coatings for corrosion protection of metals. In: A. S. H. Makhlouf, ed., *Handbook of Smart Coatings for Materials Protection* (pp. 328–369). Cambridge, UK: Woodhead Publishing Ltd.

Zhao, H., Price, W. E., & Wallace, G. G. (1998). Synthesis, characterisation and transport properties of layered conducting electroactive polypyrrole membranes. *J. Membr. Sci.*, *148*, 161–172.

2 Principles of Electrochemical Corrosion

2.1 INTRODUCTION

Corrosion of a metal or an alloy is an electrochemical reaction, in the sense that it involves transfer of charge (electrons) between a metal electrode and a chemical species. This process results in a decrease in free energy, and a change in free energy (ΔG) may be related to the potential of the corroding metal by the equation,

$$\Delta G = -nEF \qquad (2.1)$$

where n is the number of electrons involved in the reaction, F is the Faraday constant (96,500 C/mol), and E is the electrode potential of the corroding metal/metal ion or salt system. It is caused mostly by heterogeneity on a metallic surface because of lower oxygen concentration or stresses or contact between dissimilar metals. It is essential, therefore, to understand the electrochemistry of corrosion and electrode kinetics. The basic concepts of electrochemical corrosion are discussed in this chapter. However, complex mathematical treatment is avoided and the concepts are presented in a manner to serve the purpose of this book.

2.2 ELECTRODE REACTIONS AND EQUILIBRIUM POTENTIAL

In the metal lattice, the electrons are free to move and the positively charged ions occupy the fixed positions in the lattice, whereas the electrolyte contains no electrons but only ions generally solvated or hydrated, which are free to move. When the metal electrode is immersed in the aqueous electrolyte, positively charged metal ions tend to go into the solution, leaving electrons behind on the metal electrode as per expression:

$$M - M^{n+} + ne^- \qquad (2.2)$$

This is an anodic reaction that results in the oxidation of the metal atoms to its ions. The accumulation of negative charge on the metal electrode leads to an increase in the potential difference between the metal and the electrolyte. This potential difference is known as the potential of the metal. Dissolved metal ions have tendency to get deposited on the metallic surface.

$$M^{n+} + ne^- - M \qquad (2.3)$$

This is a cathodic reaction that results in the reduction of the metal ions in the solution and their subsequent deposition on the metal surface. Continued dissolution and deposition will lead to establishment of an equilibrium at which the rate of anodic and cathodic reaction becomes equal, i.e.,

$$M = M^{n+} + ne^-. \tag{2.4}$$

The metal would be able to achieve a stable potential known as equilibrium potential. It is given by a simple form of the Nernst equation (Angal, 2010; *ASM Metals Hand Book*, 2003):

$$E = E^0 \pm \frac{RT}{nF} \ln a_{M^{n+}}, \tag{2.5}$$

where E^0 is the standard electrode potential for unit activity of dissolved metal ions, $a_{M^{n+}}$, R is the gas constant (8.31 J/K/mol), T is the temperature in Kelvin, n is the number of electrons transferred per ion, and F is the Faraday constant (96,500 C/mol). In Equation 2.4, the plus sign refers to reduction and negative sign refers to oxidation potential. The Nernst equation shows that the equilibrium potential depends on the standard electrode potential, ionic activity, and temperature. The tendency of a metal to form metal ions in the solution is revealed by its standard electrode potential for the metal–metal ion reaction. Hence, the more negative value of the standard electrode potential, the higher the tendency of the metal to corrode. It is to be noted that the standard potentials have been defined as reduction potentials by international agreement. Each metal has its own standard reduction potential as shown in Figure 2.1 (Ashby et al., 2007).

Once the metal electrode potential attains the equilibrium potential, no subsequent loss of metal ions into the solution occurs. The potential of a metal electrode in an aqueous electrolyte, however, does not attain the equilibrium potential but remains more positive because electrons get consumed in cathodic reactions. The nature of the reaction at the cathode depends upon the type of the environment. In acid solutions, for example, electrons liberated during anodic reaction react with hydrogen ions adsorbed on the metal surface to generate hydrogen.

$$2H^+ + 2e^- - H_2 \tag{2.6}$$

As the hydrogen is evolved and the electrons are utilized, the reaction promotes anodic reaction to generate more and more metallic ions and electrons. Consequently, the potential of the metal electrode becomes more negative than its equilibrium potential and the metal electrode dissolution into the solution continues. If the acidic solution contains an oxidizing agent, the cathode reaction becomes oxygen reduction:

$$O_2 + 4H^+ + 4e^- - 2H_2O. \tag{2.7}$$

In neutral or basic solutions, electrons react with oxygen adsorbed on the metal surface to produce hydroxyl ions:

$$O_2 + H_2O + 2e^- - 2(OH)^-. \tag{2.8}$$

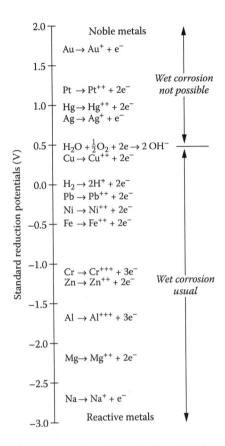

FIGURE 2.1 Standard reduction potentials of metals. (Reprinted from *Materials Engineering, Science, Processing and Design*, Ashby, M., Shercliff, H., and Cebon, D., p. 399, Copyright 2007, with permission from Elsevier.)

As the electrons are consumed, the reaction encourages anodic reaction to liberate more and more metallic ions and electrons. Consequently, the potential of the metal becomes, again, more negative than its equilibrium potential and the metal dissolution into the solution gets continued. Hydrogen evolution is common electrochemical reaction since acidic media are frequently encountered. Oxygen reduction is also common since any aqueous solution in contact with air is capable of encouraging this reaction. Metal ion reduction and deposition are less common cathodic reactions (Deshpande et al., 2014; Fontana, 2005). In a large number of electrochemical situations, the anodic reaction gets associated with more than one cathodic reaction. For example, in oxygenated acidic solution, the corrosion reaction (Equation 2.1) could be driven by both hydrogen evolution (Equation 2.5) and oxygen reduction (Equation 2.7). In case of complex metal alloys, the anodic reaction may be the sum of more than one dissolution reaction. For a smooth-single component metal surface, the anodic and cathodic sites will be separated by only a few nanometers. The anodic and cathodic sites may shift with time to give the appearance of uniform corrosion. However, atomic scale defects such as kinks and surface defects such

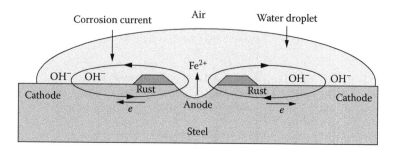

FIGURE 2.2 **(See color insert.)** Rusting of plain carbon steel in the presence of a water droplet. (With kind permission from Springer Science+Business Media: *J. Coat. Technol. Res.*, Conducting polymers for corrosion protection: a review, 11(4), 2014, p. 475, Deshpande, P. P., Jadhav, N. G., Gelling, V. J., and Sazou, D. © American Coatings Association 2014.)

as grain boundaries can lead to stabilization of discrete anodic and cathodic sites (*ASM Metals Hand Book*, 2003). In such cases, the corrosion will be recognized by the anodic sites only, and localized corrosion is said to occur. It must be noted that the combination of a large cathodic area with a small anode is always harmful. For example, when galvanized steel is painted, an unfavorable anode–cathode ratio exists at the cut edge. This leads to blistering and delamination of the paint coating.

The most common example of wet corrosion is rusting of plain carbon steel in the presence of moisture, which is depicted in Figure 2.2 (Deshpande et al., 2014). It can be explained in terms of the overall corrosion reaction as follows:

$$4Fe + 3O_2 + 2H_2O - 2Fe_2O_3xH_2O \tag{2.9}$$

The area under the rust deposit, being oxygen deficient, becomes anodic. In addition, in the rust deposit, being permeable to air and water, corrosion continues below the layers of rust deposit. To sum up, wet corrosion is a complex process that can take place in a wide variety of forms and is affected by chemical, electrochemical, and metallurgical factors such as the composition and metallurgical properties of metal or alloy, the chemical composition and physical properties of the environment, the presence or absence of surface films, and the properties of surface films (*ASM Metals Hand Book*, 2003).

2.2.1 Metal Dissolution, Charge Transfer, and Mass Transport

During metal dissolution or corrosion, electron transfer reactions occur across the interface, which leads to charge transfer current or faradic current. Reactants are transported from the bulk of the solution to the electrode surface and products are transported from the electrode surface to the bulk of the solution by diffusion, migration, and convection, as shown in Figure 2.3. In the absence of an electric field, migration becomes negligible and the convection force disappears in stagnant conditions (Roberge, 2000).

The metal dissolution and ion formation continue until the electric double layer of such strength is established, the potential of the solvated ions is raised, the potential of positive ions in the metal is lowered, and finally, both potentials become equal.

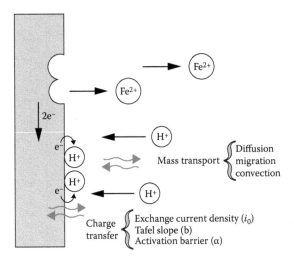

FIGURE 2.3 Charge transfer and mass transport at the interface during corrosion of plain carbon steel. (Reproduced with permission of McGraw-Hill Education from Roberge, P. R., *Handbook of Corrosion Engineering*, p. 39, New York, Copyright ©2000, The McGraw-Hill Companies, Inc.)

If the electrons are not consumed, then such a potential is developed quickly and the dissolution of ions in solution is discouraged. If the electrons are utilized, then it may take a longer period until the concentration of the metal ions in the solution becomes significant.

2.3 ELECTRIC DOUBLE LAYER

Ions and polar water molecules get attracted to the metal electrode–aqueous solution interface because of the strong electric field in the vicinity of the metal electrode. Because a water molecule is dipolar, the oxygen end of the water molecule forms the negatively charged end and the hydrogen end forms the positively charged end. If the metal electrode has excess negative charge, the water molecule gets oriented in such a way that its positive end is toward the metal electrode surface and its negative end is toward the excess positive charge in the solution. This layer of water molecules, in addition to solvent molecules containing specifically adsorbed ions on the electrode surface, form an electric double layer that is known as the Helmholtz double layer, as shown in Figure 2.4 (Bard & Faulkner, 2004).

The line drawn through the center of these molecules is known as the inner Helmholtz plane. Also, the water molecules get attracted to the charged ions in the electrolyte and align themselves in the electric field established by the charge of ion. Because the electric field is strongest close to the ion, some water molecules reside very close to an ion in the solution. The metallic ions that enter into the solution thus get hydrated or solvated. The electric field is so intense that these water molecules travel with the ion as it moves through the solvent. The line drawn through the center of such hydrated cations at a distance of closest approach is known as the outer Helmholtz plane. A potential difference that is thus created between the

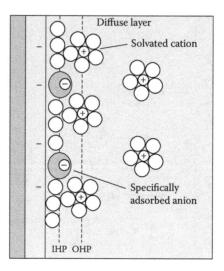

FIGURE 2.4 Helmholtz double-layer model for concentrated solution. IHP, inner Helmholtz plane; OHP, outer Helmholtz plane. (Bard, A. J. and Faulkner, L. R., *Electrochemical Methods—Fundamentals and Applications*. p. 13. 2004. Copyright Wiley-VCH Verlag GmbH & Co. KGaA. Reproduced with permission.)

metal electrode and the solution varies linearly with the distance from the metal surface. The Helmholtz model is, however, applicable to concentrated solutions only. When the solution becomes more dilute, a diffuse mobile layer known as the Gouy-Chapman layer is said to form, which extends up to 1 μm into the solution from the outer Helmholtz layer plane. The net charge of this diffuse layer is equal and opposite to that of the Helmholtz layer. As the diffuse layer contains excess cations or anions, the potential varies exponentially with the distance. A combination of these two models, known as the Stern model, takes ionic interaction in the solution into consideration while describing electric double layer. The electric double layer, when established on the metal–electrolyte interface, prevents the close approach of dissolved ions from the solution and discourage subsequent dissolution of the metal. A dynamic equilibrium is thus achieved owing to the formation of double layer, with no net flow of metal ions into the solution. Electric double layer stores the charges and works as a parallel plate capacitor. Hence, the electric equivalent of a metal–aqueous solution interface, in case of no charge transfer, can be represented by a simple capacitor. The metal–electrolyte interface is a barrier to the transfer of electrons from or to the metal: This can be represented by charge transfer resistance (R_{ct}). This resistance is not a simple ohmic one because it varies with the electrode potential. The electrical equivalent circuit of a metal–electrolyte interface, in case of the charge transfer, can be represented by a parallel combination of a double layer capacitance (C_{dl}) and a charge transfer resistance (R_{ct}), as shown in Figure 2.5.

In Figure 2.5, R_s stands for the solution resistance. A detailed discussion on the use of equivalent electrical circuits in modeling of corrosion process can be found in Chapter 5.

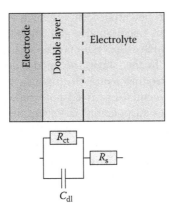

FIGURE 2.5   Electric double layer and its electrical circuit equivalent.

2.3.1   ELECTRODE KINETICS, POLARIZATION, AND MIXED POTENTIAL THEORY

When a metal electrode is not corroding, the local potentials are the open circuit potentials. The electrode potentials $E_{0(a)}$ and $E_{0(c)}$ as shown in Figure 2.6 are the open circuit potentials of the anodic and cathodic areas, respectively, on the metal surface or the two potentials of two different metals in case of galvanic coupling. The symbols $i_{0(a)}$ and $i_{0(c)}$ in Figure 2.6 are exchange current densities of the anode and cathode, respectively. The concept of exchange current density is elaborated in the next section.

When a metal electrode begins to corrode, the potential of the cathodic areas becomes more anodic and the potential of the anodic areas becomes more cathodic. The rates of anodic reaction increase and the rates of cathodic reaction decrease with continued metal dissolution. The departure of the electrode potential from its equilibrium value or open circuit potential, known as polarization, can be quantified in terms of overvoltage or over potential as follows (Perez, 2004):

$$\eta = E - E_{ocp} \tag{2.10}$$

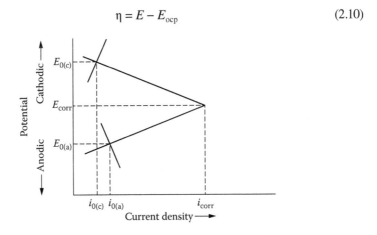

FIGURE 2.6   Stern diagram: Polarization as a function of current density.

The shift in equilibrium potential occurs in such a way that the flow of corrosion current gets reduced. Local anodes and cathodes on the surface undergo continuous polarization, which causes their potentials to change continuously until a steady state is reached. In this stage, the anodic and cathodic reaction rates become equal and the corroding metal electrode in a solution attains a potential known as corrosion potential or mixed potential, E_{corr}. The current density corresponding to the corrosion potential is known as corrosion current density since it represents the rate of metal dissolution. Wagner and Traud proposed the mixed potential theory in 1938 and laid the foundation of modern electrode kinetics theory. These researchers postulated two hypotheses: (1) Any electrochemical reaction is divisible into two or more partial reactions and (2) there can be no net accumulation of charge during an electrochemical reaction (Fraunhofer, 1974). The electrode kinetic behavior of pure iron in an acidic solution is shown in Figure 2.7 (Fontana, 2005).

The standard electrode potential of iron when in contact with a solution of ferrous irons at unit activity (that is, 1 g ion/L Fe^{2+} ions in solution) and at standard temperature and pressure conditions is -0.44 V against the standard hydrogen electrode. This is the theoretical or equilibrium potential of iron. When it is immersed in acid solution, iron dissolution and hydrogen evolution occur simultaneously, and therefore its potential cannot remain at either of the standard equilibrium potential values for the separate reactions that are occurring. Hence, the potential of an iron electrode, when immersed in acidic solution, is found to lie somewhere between those for hydrogen evolution reaction and the iron dissolution reaction, i.e., at mixed potential or corrosion potential, as shown in Figure 2.7.

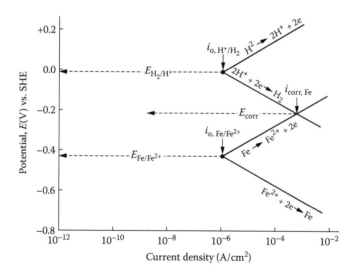

FIGURE 2.7 Electrode kinetic behavior of pure iron in acid solution. (Source: Fontana, M. G., *Corrosion Engineering, Third Edition,* Tata McGraw-Hill, 2005.)

2.3.2   Exchange Current Density and Its Significance

During dissolution of metal into the solution, the ionic species are produced and discharged simultaneously at the metallic surface. Since the ions are charged species, the ions that flow constitute two equal and opposite electric currents. Consider, for example, oxidation and reduction of hydrogen. As two electrons are consumed during the reduction of the two hydrogen ions and since two electrons are released during the oxidation of the single hydrogen molecule, the electrochemical reaction rate during oxidation and reduction of hydrogen can be expressed in terms of current density. At equilibrium, the rate of oxidation and reduction reactions must be equal:

$$r_{oxid} = r_{red} = \frac{i_0}{nF}, \qquad (2.11)$$

where i_0 is the exchange current density, n is the number of electrons consumed, and F is the Faraday constant. It gives the magnitude of currents leaving and entering the metal in the solution at equilibrium and therefore plays an important role in determining the rate of corrosion. For example, the standard electrode potentials for iron and zinc are −0.44 V and −0.76 V, respectively. Although zinc is more active, it corrodes more slowly than iron does in hydrochloric acid. This is because of the very low exchange current density for the hydrogen evolution reaction on zinc. The exchange current density of zinc dissolution is 10^{-7} A/cm^2, whereas for the iron, it is 10^{-6} A/cm^2 at −0.44 V. This principle is used in corrosion protection of steel using galvanizing coating as the zinc begins to corrode first in case of damage of coating, if any, and it corrodes much more slowly than the steel (Fontana, 2005).

2.3.3   Causes of Polarization

Accumulation of electrons coming from the anode causes polarization of the cathode and results in the shift of potential in a negative direction. This effect, known as activation polarization, can be illustrated with the help of the hydrogen reduction reaction at the cathode as shown in Figure 2.8.

The steps in the formation of hydrogen gas at the cathode are as follows: migration of hydrogen ions to the surface, flow of electrons to hydrogen ions, formation of atomic hydrogen and formation of diatomic hydrogen molecules, and the formation of hydrogen gas. Any of these steps can be a rate-limiting step and therefore cause the polarization of the metal electrode. As the metal dissolution proceeds, the ionic concentration in the vicinity of each electrode becomes different from that of the electrolyte. At the anode, concentration of the dissolved metal ions will substantially increase, making the anode more positive; at the cathode, the concentration of the positive ions will decrease, making the cathode potential more negative. In this case, the displacement of electrons away from the anode through the metal is faster than the electrode processes such as transfer of metal ions by the diffusion and convection. This results in an excess of positive charges on the metal and, thereby, a shift of the potential in the positive direction. This phenomenon, known as concentration

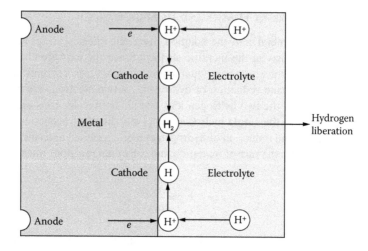

FIGURE 2.8   Activation polarization.

polarization, can be explained by considering the hydrogen evolution reduction reaction. When the reaction rate is low and or the concentration of H^+ ions is high, there is always an adequate supply of hydrogen ions available in the solution at the region near the electrode interface as shown in Figure 2.9a. On the other hand, at high rates and or low H^+ concentrations, ions from the solution are unable to diffuse across the double layer quickly. As such, a depletion zone may be formed in the vicinity of the interface as shown in Figure 2.9b. Thus, the diffusion of H^+ ions at the interface is rate controlling and the system is said to be concentration polarized.

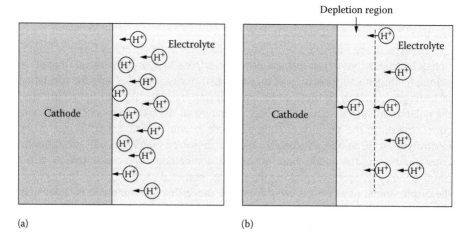

(a)   (b)

FIGURE 2.9   (a) Concentration polarization and (b) depletion zone.

It can be said that the concentration polarization is controlled by diffusion gradients, whereas a kinetic factor such as charge transfer is responsible for activation polarization. Resistance polarization of the electrode occurs because of the resistance to the passage of ions in the solution, i.e, IR drop. The total polarization, therefore, can be written as follows:

$$\eta = \eta_a + \eta_c + \eta_r. \tag{2.12}$$

In addition to these effects, the metal electrode may get polarized by the formation of a passive film as shown in Figure 2.10.

Initially, the metal is at equilibrium and its exchange current density is i_0. As the electrode potential is made more positive, the metal behaves as active metal. Its dissolution rate increases exponentially. When the potential becomes more positive and reaches the primary passivation potential, the current density and, hence, corrosion rate decrease. At this potential, the metal forms a protective film on its surface, which is responsible for preventing subsequent dissolution. E_p is the potential above which the system becomes passive and exhibits passive current density i_p. The critical current density for passivation is i_c. As the potential is made more positive, the current density remains the same over the passive region. A still further increase in potential beyond the passive region destroys the passive film and makes the metal more active again in the trans-passive region. A passivable metal exhibits the characteristic S-shaped dissolution curve. A number of metals and alloys such as stainless steel exhibit this kind of behavior, which is known as passivity. Passivity can be employed to protect the metals from corrosion. Surface treatment can be carried out on an alloy capable of being passivated or the environment can be altered to generate

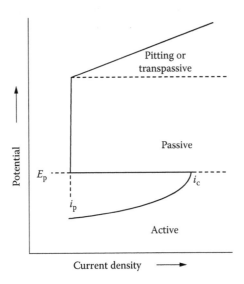

FIGURE 2.10 Polarization of passive metal. (From *ASM Metals Handbook Vol 13 A— Corrosion: Fundamentals, Testing and Protection*, p. 61, 2003. Reprinted with permission of ASM International. All rights reserved. http://www.asminternational.org.)

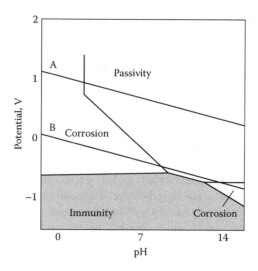

FIGURE 2.11 Simplified potential pH diagram for the iron water system. (Adapted from *ASM Metals Handbook, volume 13 A: Corrosion: Fundamentals, Testing and Protection,* 2003.)

passive film on the metallic surface. Passivation of a metal by anodic polarization and oxide film formation is known as anodic protection. Conditions of passivity can be determined by using the potential pH diagram, the Poubaix diagram. Figure 2.11 shows a simplified version of the pourbaix diagram for the iron water system. The three theoretical regions show, on a thermodynamic basis, the potential pH conditions where no corrosion is anticipated (immunity), where a corrosion product film that may provide corrosion protection (passivation), and where corrosion is expected (corrosion).

Above equilibrium line A, oxygen is evolved, and below equilibrium line B, hydrogen is generated. The potentials above the oxygen evolution line are in the trans-passive region. The diagrams can therefore be used in identifying the active, passive, and trans-passive regions on the active passive polarization curves (*ASM Metals Hand Book*, 2003).

REFERENCES

Angal, R. D. (2010). *Principles and Prevention of Corrosion*. New Delhi, India: Narosa Publishing House Pvt. Ltd.

Ashby, M., Shercliff, H., & Cebon, D. (2007). *Materials Engineering, Science, Processing and Design* (p. 399). Oxford, UK: Elsevier.

ASM Metals Hand Book, vol. 13, A—Corrosion: Fundamentals, Testing and Protection. (2003). Materials Park, OH: ASM International.

Bard, A. J., & Faulkner, L. R. (2004). *Electrochemical Methods—Fundamentals and Applications*, 2nd ed. (p. 13). New Delhi, India: Wiley India Pvt. Ltd.

Deshpande, P. P., Jadhav, N. G., Gelling, V. J., & Sazou, D. (2014). Conducting polymers for corrosion protection: A review. *J. Coat. Technol. Res., 11*(4), 475, doi:10.1007/s 11998-014-9586-7.

Fontana, M. G. (2005). *Corrosion Engineering*, 3rd ed. New Delhi, India: Tata McGraw-Hill.
Fraunhofer, J. A. (1974). *Concise Corrosion Science*. London, UK: Portcullis Press Ltd.
Perez, N. (2004). *Electrochemistry and Corrosion Science*. New Delhi, India: Springer (India) Pvt. Ltd.
Roberge, P. R. (2000). *Hand Book of Corrosion Engineering*. New York: McGraw-Hill.

3 Corrosion Prevention Mechanisms

3.1 INTRODUCTION

The elucidation of corrosion protection mechanisms is rather complex by the number of process variables involved, such as the type of conducting polymer, the method of synthesis, the presence or absence of a top coat, defects in a top coat if the top coat is used, the type of metal substrate and its pretreatment, the nature of the corrosive environment, and test method. Consequently, a number of hypotheses have been proposed from time to time and described in the next paragraphs. Tallman et al. (2002a) published a comprehensive review paper in which the literature is interpreted in terms of the proposed corrosion protection mechanisms. Understanding these mechanisms is essential from both fundamental and application points of view.

3.2 ACTIVE ELECTRONIC BARRIER AT THE METALLIC SURFACE

When a contact is established between a metal and a semiconductor, known as a Schottky junction, electrons will flow from the material having the highest Fermi level to that with the lowest Fermi level to make the levels equal. A charge is built up on both sides of the interface and because the semiconductor has a low charge carrier density that will result in a positively charged film known as the space charge layer. A metallic surface thus holds a positive dipole layer when it is in contact with a doped semiconductor or conducting polymer. This interfacial space charge layer generates an electric field. Consequently, bending of the semiconductor band occurs at the metal–semiconductor interface. The net band bending is defined as the active electronic barrier. This active barrier impedes the flow of electrons from the metal to the oxidizing species in the environment and thereby prevents corrosion. A schematic illustration of the electron transfer during oxidation and built-in electric field at the metal–semiconductor interface is shown in Figures 3.1 and 3.2.

This novel approach to corrosion protection was demonstrated by Jain et al. (1986) using Al-indium tin oxide (ITO) and Al-SiO$_2$-ITO samples and anticipated that the concept of an active barrier could be applied to semiconducting coating such as doped polyacetylene on metal. An important feature of this mechanism is that the electric field created is intrinsic and extends up to 250 µm in dimension, providing protection.

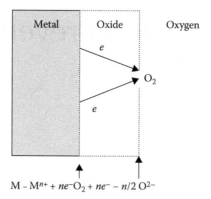

FIGURE 3.1   Electron transfer during oxidation.

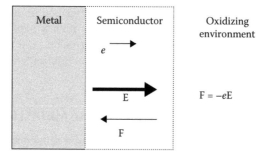

FIGURE 3.2   Built-in electric field at the metal–semiconductor interface.

3.3 BARRIER PROTECTION

Electrochemically undoped conducting polymer, i.e., conducting polymer in its insulating state, is anticipated to work as a conventional barrier paint coating that allows penetration of water and oxygen but restricts the access of corrosive species to the metal surface. Beck (1988) studied electrochemically deposited polyaniline, polypyrrole, and polythiophene coatings on iron. Corrosion prevention was observed in case of the coatings having thickness more than 1 μm, and this effect was attributed to a barrier protection (Beck, 1988). Wessling (1994) observed that the corrosion current of steel coated with polyaniline (Versicon) was lower than that of uncoated steel and that the reduction in current was greater for thicker polyaniline coatings. Sufficient thickness of the polyaniline layer was therefore considered to be essential for protection in this work (Wessling, 1994). Numerous studies, however, have indicated that conducting polymer coatings are not a simple barrier but offer active protection. Aahmad and MacDiarmid (1996), for example, observed a significant increase of +550 mV in corrosion potential for emeraldine-base-coated stainless steel. It was concluded that a conducting polymer with a higher open circuit potential than the minimum passivation potential for stainless steel in corrosive medium can provide excellent protection (Aahmad & MacDiarmid, 1996). The experiment in this work was based on the assumption that the conducting polymer can provide enough current density (micro to milli amperes per square centimeter) that is required for anodic protection. Cathodic protection needs a higher current density that cannot be supplied by the conducting polymer. The stronger evidence that polyaniline coatings are not simple barriers is given in those studies where corrosion protection was observed when an intentional defect was made in the coating. Darowicki and Kawula (2005), for example, investigated the effect of electrodeposited polyaniline primer on the anticorrosion properties of epoxy coating on low carbon steel. To stimulate defected coating, four coatings were damaged by drilling a 0.8-mm hole in the coating system. Corrosion protection was observed even after an artificial defect was introduced into the conducting polyaniline coating (Darowicki & Kawula, 2005). If the protection mechanism depends entirely on the barrier properties, damage tolerance cannot be expected. It can be said, therefore, that mechanisms in addition to or other than simple barrier protection are certainly working toward corrosion protection by conducting polymer coatings.

3.4 SELF-HEALING

Organic coating generally gets damaged during its use. In this situation, a self-healing of corrosion defect on the metallic surface might be initiated depending on the nature of the metal and doping anions. In a number of metals and or alloys, such as steel and aluminum, a passive oxide layer is generated on the metallic surface (anodic protection mechanism) or the doping anions that are released from the coating act as inhibitors (controlled inhibitor release mechanism).

3.4.1 ANODIC PROTECTION

The shift of the potential in anodic direction is known as ennobling. Conducting polymers such as polyaniline is able to ennoble the surface of iron or steel since its redox potential is in the range 0.4 to 1.0 V (vs. standard hydrogen electrode (SHE) at pH 7), which is substantially higher than that of iron (-0.62 V vs. SHE at pH 7) or steel (Tallman et al., 2002a). A number of studies have shown a shift in corrosion potential in the positive direction in case of steel samples coated with conducting polyaniline and conducting polypyrrole (Tallman et al., 2002b). Deshpande et al. (2012), however, noted a decrease in corrosion potential of uncoated steel from -680 mV to -743 mV in case of 2 wt% polyaniline-based epoxy painted steel. It must be noted here that in this work, stainless steel was used as the counter electrode. It can be concluded from these results that conducting polyaniline and conducting polypyrrole coatings are capable of significant ennobling of steel. However, a limitation of ennobling is that if the coating is damaged, protection will cease to act since the underlying exposed metal would serve as anode due to its lower potential. Thus, ennobling is essential but not sufficient for the corrosion protection of metals and alloys. In association with shift in potential, there must be a formation of passive oxide layer on metallic surface to achieve efficient protection, a technique known as anodic protection. A simplified Pourbaix diagram to explain this approach is shown in Figure 3.3.

Berry (1985) found that electrochemically deposited polyaniline film reduced corrosion rate of 410 and 430 ferritic stainless steel in sulfuric acid and noted that the

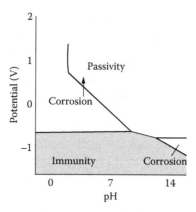

FIGURE 3.3   Anodic protection of steel.

electrochemical deposition of polyaniline was preceded by the formation of a passive oxide layer on the steel surface. The protective mechanism was then thought to be anodic protection that maintains the passive oxide film on the stainless steel. A number of investigators have tried to prove anodic protection mechanism either by electrochemical methods or by surface analysis or by using both approaches. Wessling (1996) and Lu et al. (1995) demonstrated that when doped polyaniline is placed in contact with steel, the steel surface undergoes rapid oxidation to provide a layer of γ-Fe_2O_3 at the polyaniline–iron interface. Passivation reaction sequence, due to redox catalytic action of polyaniline, leading to the formation of oxide layer is shown in Figure 3.4 (Wessling, 1996).

Auger and X-ray photoelectron spectroscopy (XPS) studies revealed that the oxide layer is composed of γ-Fe_2O_3 on the outside with an Fe_3O_4 layer sandwiched between the γ-Fe_2O_3 layer and iron. The layer of γ-Fe_2O_3 formed at the interface is dense and provides barrier protection. It is important to note that during the formation of passive iron oxide layer, the iron is oxidized and polyaniline gets reduced. The reduced polyaniline can be reoxidized to the emeraldine form, as shown in the passivation scheme. This transformation imparts a self-healing property to the polyaniline if the coating is damaged. The generalized mechanism for corrosion protection of the exposed steel surface adjacent to the polyaniline as a primer and epoxy as a top coat is shown in Figure 3.5.

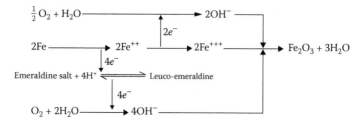

FIGURE 3.4   Passivation scheme. Wessling, B.: Corrosion prevention with an organic metal (polyaniline): Surface ennobling, passivation, corrosion test results. *Materials and Corrosion.* 1996. 47: 443. 1996. Copyright Wiley-VCH Verlag GmbH & Co. KGaA. Reproduced with permission.

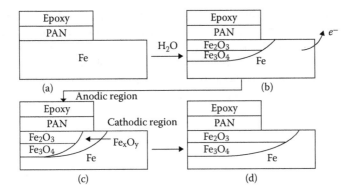

FIGURE 3.5 Corrosion protection mechanism by conducting polyaniline coating as a primer and epoxy as a top coat. (From Lu, W. et al., *Handbook of Conducting Polymers*, second edition (ed. Skothem), Marcel Dekker, New York, USA, p. 918, 1998.)

As stated earlier, when doped polyaniline comes in contact with the iron surface, a layer of γ-Fe_2O_3 is formed at the interface (a to b). The sample comes in contact with the corrosive environment if the epoxy coating is damaged. In this situation, the bare steel surface is cathodically protected and the reduced polyaniline becomes the anode. The reduction reaction occurs on the bare steel and protects it from oxidation (cathodic protection). This mechanism continues as long as enough polyaniline is present in its reduced state relative to its emeraldine form. Ultimately, all of the reduced polyaniline will be reoxidized to the emeraldine form (b to c). This process cannot continue for a longer time and stops after a short time. The region on the bare steel surface close to the polyaniline layer becomes anodically polarized because of its close proximity with the high potential polyaniline. This region, being anode, undergoes oxidation reaction. The region, which is away from the conducting polymer edge, becomes the cathode where reduction reactions occur. This localization of anodic and cathodic reactions is responsible for the gradual growth of the passivating oxide layer on the metal surface (c to d). The distance of the passive oxide layer from the edge of the coating depends upon the oxidizing power of the conducting polymer, the electrolyte conductance, the oxygen concentration in the solution, and the pH of the medium. For example, in dilute HCl, where the oxidizing power is high, it can extend up to 6 mm from the edge of the coating, whereas it can extend only up to 1–2 mm from the edge of the coating in dilute neutral NaCl solution. It can be argued, therefore, that the conducting polyaniline coating shifts corrosion potential of the metal to the passive region so that a protective oxide layer forms on the surface and prevents subsequent corrosion even to damaged areas in the coating where small areas of steel are exposed (Lu et al., 1998). Kinlen et al. (1997) offered a simple explanation of the anodic protection mechanism, according to following equation:

$$1/n\,M + 1/m\,\text{PAN} - ES^{m+} + y/n\,H_2O — 1/n\,M(OH)_y^{(n-y)+}$$
$$+ 1/m\,\text{PAN} - ES^0 + y/n\,H^+$$

Thus, the conducting polyaniline shifts the corrosion potential of the metal to the passive region, so that a protective oxide layer forms on the metal. The reduction of oxygen to hydroxide shifts from the metallic surface to the polymer–electrolyte interface and probably involves the reoxidation of the polymer, which in turn stabilizes the polymer from cathodic disbonding, as given in the following expressions (Kinlen et al., 1997; Tallman et al., 2002a):

$$1/4 \; m \; O_2 + 1/2 \; m \; H_2O + me^- \longrightarrow m \; OH^-$$

$$1/4 \; m \; O_2 + 1/2 \; m \; H_2O + PAN - ES^0 \longrightarrow PAN - EB^{m+} + m \; OH^-$$

In the process, metal is oxidized, conducting polymer gets reduced, and finally, it can be reoxidized by atmosphere or dissolved oxygen from the electrolyte. The anodic protection mechanism, however, cannot be used to understand corrosion inhibition of steel by undoped polyaniline in 3.5% NaCl. Moreover, this coating system protects the metal even in case of damage of the coating (Lu et al., 1998). However, a mechanism is also proposed which assigns the effect of inhibition by dopant ions released upon reduction of the polyaniline emeraldine salt to its non conducting leuco counterpart (Cook et al., 2004). It is established that metals or alloys can be anodically protected if they exhibit active–passive behavior. Moreover, passivation will occur only under free corrosion conditions if the corrosion potential exceeds the passivation potential. Hence, the theory of anodic protection may be employed to explain corrosion protection of stainless steel by conducting polyaniline. Sathiyanarayanan et al. (2005) proposed that because of the conducting nature of the polyaniline coating, the oxygen reduction reaction takes place on the coating while the oxidation of ferrous ions to iron oxides takes place on the exposed iron surface at pinhole areas and under the coating in neutral media. In acid media, the passivation of pinholes takes place by cathodic complimentary conversion of emeraldine salt to leuco-emeraldine, as shown in Figure 3.6.

In another attempt, Schaur et al. (1998) investigated corrosion protection of iron with commercial polyaniline primers (supplied by Ormecon Chemie, Ahrensburg, Germany) to study the protective action of polyaniline as well as to understand the factors influencing the protective efficiency of the primers. In this work, it is argued that the corrosion protection by conducting polyaniline primer takes place in three stages:

1. First stage: The interaction between polyaniline and the substrate initiates passivation due to the formation of Fe_3O_4 and γ-Fe_2O_3 iron oxides and provides the basis for active corrosion protection by polyaniline in the second stage.
2. Second stage: As soon as water and corrosive ions diffuse through the coating and reach the metal substrate, this active corrosion protection action begins, as shown in Figure 3.7.

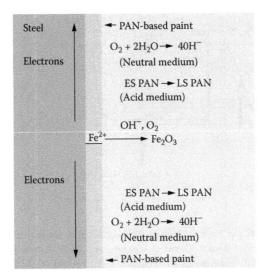

FIGURE 3.6 Passivation mechanism. (Reprinted from *Progress in Organic Coatings*, 53, S. Sathiyanarayanan, S. Muthukrishnan, G. Venkatachari, and D.C. Trivedi, Corrosion protection of steel by polyaniline (PANI) pigmented paint coating, 301, Copyright 2005, with permission from Elsevier.)

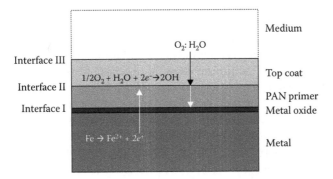

FIGURE 3.7 Active corrosion protection by conducting polyaniline primer as a result of separation of anodic and cathodic reaction. (Reprinted from *Prog. Org. Coat.*, 33, Schauer, T., Joos, A., Dulog, L., and Eisenbach, C., Protection of iron against t corrosion with polyaniline primers, p. 25, Copyright 1998, with permission from Elsevier.)

Electrons from the anodic partial reaction are transported by polyaniline to interface between the polyaniline primer and the top coat and react there with oxygen to form hydroxyl ions as shown by the following equations:

$$Fe - Fe^{2+} + 2e^-$$

$$1/2 O_2 + H_2O + 2e^- - 2OH^-$$

The oxygen concentration gradient across the coating and the limited supply of electrons at the metal surface result in the separation of partial cathodic and anodic corrosion processes.

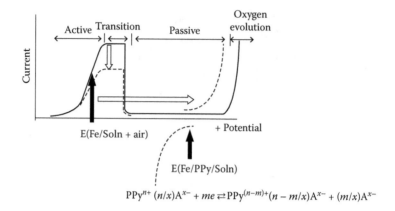

$$PPy^{n+} (n/x)A^{x-} + me \rightleftharpoons PPy^{(n-m)+}(n-m/x)A^{x-} + (m/x)A^{x-}$$

FIGURE 3.8 Potential current relation of steels covered by oxidative conducting polypyrrole. (Reproduced from Ohsuka, T., *Int. J. Corros.*, vol. 2012, Article ID 915090, p. 4, doi: 1155/2012/915090. With permission. Copyright © 2012 Toshiaki Ohtsuka.)

3. Third stage: Polyaniline emeraldine salt reacts with OH⁻ ions to form emeraldine base and thereby limits the increase in pH at the primer–metal interface. Thus, conditions at the metal surface become more favorable for the maintenance of a passive state and formation of stable oxides. The transition from emeraldine salt to emeraldine base imparts barrier protection in this stage (Schauer et al., 1998). No reversible transformation, even by changing pH and oxygen concentration, was found in this work. Therefore, it was concluded that phase transition processes did not contribute in the corrosion protection. Consequently, this mechanism cannot be extended to explain corrosion protection offered by the emeraldine base form of the polyaniline. More investigations, therefore, are necessary to confirm separation of anodic and cathodic reactions. In case of polypyrrole coating on steel, it is suggested that the barrier effect of polypyrrole suppresses active dissolution of the steel and oxidative property of polypyrrole shifts the potential into the passive state (Ohtsuka, 2012). In solution at neutral pH, the open circuit potential of uncoated steel lies in the active potential region and the corrosion rate of the steel is usually high. Owing to the coating of conducting polymer, the maximum current in the active–passive transition region was limited by the barrier effect and then the potential can be shifted to the higher potential in the passive state by the strongly oxidative property of the conducting polymer, as shown in Figure 3.8.

A major problem in anodic protection is the possibility of a breakdown of passive film as a result of attack of aggressive ions such as chloride ions in the solution and subsequent corrosion.

3.4.2 CONTROLLED INHIBITOR RELEASE

When a scratch exists on the conducting-polymer-coated metal surface, there is a galvanic coupling between the metal and the conducting polymer, as shown in Figure 3.9.

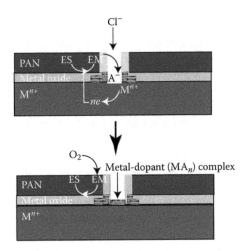

FIGURE 3.9 (See color insert.) Controlled inhibitor release mechanism. (With kind permission from Springer Science+Business Media: *J. Coat. Technol. Res.*, Conducting polymers for corrosion protection: a review, 11(4), 2014, p. 482, Deshpande, P. P., Jadhav, N. G., Gelling, V. J., and Sazou, D. © American Coatings Association 2014.)

The coating becomes cathodic to the metal scratch, which is anodic. The anodic corrosion reaction drives reduction of the coating as it exists in a doped acidified state. Kendig and Warren (2003) anticipated that doping anions with corrosion-inhibiting properties inside the polymer matrix gets released during reduction of the polymer coating and migrate to the corroding defect. Here, the inhibitor anion could significantly decrease the corrosion rate (Kendig & Warren, 2003). Selection of an anion dopant is therefore critical for corrosion protection and may be metal sensitive. The conducive polymer would act as storage for corrosion inhibitors, which supplies them immediately just after corroding defect appears. The efficacy of the inhibition is strongly dependent on the concentration of the inhibitor. The transport of the inhibitor must be fast enough to ensure proper concentration. Controlled inhibitor mechanism therefore depends on the ionic conductivity of the coating and its ability to release doping anions from the polymer matrix. It is not clear whether or not the inhibiting anions should be anodic or cathodic inhibitors of the oxygen reduction reaction. Very little work has been carried out to study the effect of dopant anion available in conducting polyaniline on the corrosion protection. Nevertheless, Kinlen et al. (2002) observed that phosphanate-type dopants were better than sulfonic acid dopants. The same team demonstrated that the exposed area due to an artificial hole in the polyaniline coating on plain carbon steel was protected by the formation of an insoluble salt; therefore, the corrosion proceeds with an active iron (Kinlen et al., 1999). However, recent findings suggest that conducting-polymer-based coatings with an extended percolation network will fail to provide protection when relatively large defects exist since a fast breakdown of the entire coating is possible by rapid reduction of the conducting polymer induced by cation transport. In this work, further, it is argued that not high conductivity, which requires extended percolation networks, but electrochemical activity is important for corrosion protection (Rohwerder & Michalik, 2007).

3.5 CATHODIC PROTECTION

Yan et al. (2009) found that neutral or n-doped poly(2,3-dihexylthieno[3,4-b]pyr-azine) can provide cathodic protection to aluminum alloy. But for the neutral poly-pyrrole, cathodic protection did not happen (Yan et al., 2009). Recently, Li et al. (2011) studied the anticorrosion performance of emeraldine base coating on mild steel having artificial defect exposed to 1 wt% NaCl solution. In this work, it is argued that conventional mechanisms are not useful to explain good anticorrosion performance of emeraldine base on the exposed mild steel since it cannot be pas-sivated in brine, there are no dopant anions released on reduction of emeraldine base to leuco-emeraldine base, and the cathodic protection does not work in corro-sion protection, rather dissolved oxygen diffusion caused by emeraldine base plays an important role (Li et al., 2011). Recently, Elkais et al. (2013) demonstrated the use of benzoate-doped polyaniline coatings in cathodic protection and suggested the "switching zone" mechanism. Dung Nguyen et al. (2004) tried to establish the mechanism of corrosion protection using polyaniline and polypyrrole by measuring local potential, current, and impedance spectra around the defect area of the coating. It was found that the conducting polymer was able to passivate a defect area where the bare steel surface was directly in contact with the electrolyte. However, when the defect area was large, there was not sufficient charge, and the ohmic drop between the bare area and the coated area obstructed the passivation of the exposed area. In this work, it was stated that both mechanisms of corrosion protection could be work-ing simultaneously, and the various degrees of oxidation coexisted in polyaniline and participated both in the cathodic process and anodic passivation of metal (Dung Nguyen et al., 2004).

3.6 DISLOCATION OF OXYGEN REDUCTION

Electrons generated during metal oxidation at the defect area of the polymer coat-ing can go into the polymer and dislocate the oxygen reduction process from the metal–polymer interface. This would hinder coating disbondment caused by inter-facial oxygen radicals and or hydroxide ions (Tallman et al., 2002a). Rammelt et al. (2001) obtained ultrathin films (thickness ~1 μm) of polymethylthiophene on mild steel. These films were found to effectively separate the electrochemical processes of oxygen reduction and iron dissolution in the surface region. The electronic conduc-tivity of the coating is therefore important for the dislocation of oxygen reduction.

3.7 CORROSION INHIBITION

Corrosion inhibition refers to the adsorption of monomers or conducting polymers on the surface that endorses adhesion. Similar to barrier coatings, the adsorbed organic molecules can oppose anodic and cathodic reactions, i.e., electron transfer, and thereby reduce the rate of corrosion reactions. The extent of corrosion inhibition by inhibitor depends upon its ability to get adsorbed on the metallic surface, which consists of the replacement of water molecules by soluble organic species at a corrod-ing interface as given by the following reaction (Santos et al., 1998):

$$\text{Org}_{\text{sol}} + n\,\text{H}_2\text{O}_{\text{ads}} - \text{Org}_{\text{ads}} + n\,\text{H}_2\text{O}_{\text{sol}}.$$

The most powerful inhibitors are organic compounds having π-bonds in their structure. Many reports are published on the use of aniline, pyrrole, and their derivatives for corrosion protection of metals in different electrolytes. Nathan (1953), for example, reported the use of various aliphatic and aromatic amines as corrosion inhibitors in hydrochloric acid. In this work, it was observed that the efficiency of inhibition increased with the molecular weight of the inhibitor and the adsorbed film of inhibitor on the metallic surface precluded metal dissolution (Nathan, 1953). Sathiyanarayanan et al. (1992) found that ortho substituted water-soluble poly(ethoxyaniline) inhibits the corrosion of iron in 1 N HCl solution. In this work, it was found that the poly(ethoxyaniline) system was able to provide more than 80% corrosion efficiency even with 25 ppm as compared with the monomers. The authors argued that the coexistence of delocalized π-electrons and the quaternary ammonium nitrogen in the polymer promotes strong adsorption on the iron surface, which results in good corrosion inhibition. Double-layer capacitance investigations indicated that there was strong adsorption of the polymer to the metal surface. These studies also revealed that an increase in inhibitor concentration in electrolytes can increase the charge transfer resistance of the surface and thereby increase corrosion inhibition. Thus, there is a strong tendency for aniline, its derivatives, and polyaniline to adsorb on the metal surface. Consequent adhesion of the organic molecules to the metallic surface reduces the rate of the corrosion reactions. Iroh and Su (2000) suggested that the double bonds and the polar–NH group in the ring causes strong adsorption of polypyrrole and protects the underlying metal. It can be concluded, therefore, that the corrosion inhibition by adsorption of monomers and or conducting polymers is an important corrosion protection mechanism.

3.8   KRSTAJIC'S MODEL

Krstajic et al. (1997) investigated corrosion protection offered by polypyrrole coating on mild steel in sulfuric acid. These authors observed ennobling of the sample only for short duration and therefore argued that polypyrrole coatings cannot provide anodic protection to mild steel. Electrochemical methods indicated that dissolution of steel occurred at the bottom of pores in the polypyrrole coating as a result of the reduction in polypyrrole films around the pores (instead of reduction of hydrogen in case of uncoated steel). Diffusion of ions through the coating became progressively difficult with the polypyrrole reduction; consequently, corrosion rate decreased.

3.9   REDUCTION–OXIDATION

Sathiyanarayanan et al. (2006) used polydiphenylamine in vinyal coating to protect steel in 3% NaCl (Figure 3.10). The coating containing 3% polydiphenylamine is found to protect the steel effectively. During the oxidation of iron to iron oxide, the polydiphenylamine is reduced. The reduced form of polydiphenylamine is reoxidized by dissolved oxygen reduction reaction, as shown in Figure 3.10.

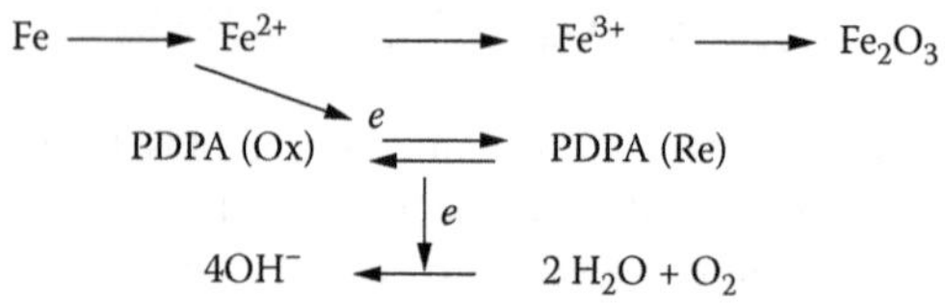

$$Fe \longrightarrow Fe^{2+} \longrightarrow Fe^{3+} \longrightarrow Fe_2O_3$$

$$PDPA\ (Ox) \overset{e}{\rightleftharpoons} PDPA\ (Re)$$

$$4OH^- \overset{e}{\longleftarrow} 2H_2O + O_2$$

FIGURE 3.10 Corrosion protection by coating containing polydiphenylamine (PDPA). (Reprinted from *Syn. Met.*, 156, Sathiyanarayanan, S., Muthukrishnan, S., and Venkatachari, G. Synthesis and anticorrosion properties of polydiphenylamine blended vinyl coatings, 1212, Copyright 2006, with permission from Elsevier.)

3.10　IRON DIFFUSION INHIBITION

Fahlman et al. (1997) suggested that when emeraldine base is coated on steel, electrons are given from the metal to the lowest unoccupied molecular orbital (LUMO) level of the polymer, thus producing a positively charged metal surface. Diffusion of water through the porous oxide leads to corrosion at the Fe–Fe_3O_4 oxide interface. Ferrous ions diffuse along grain boundaries or through vacancy hopping. At the Fe_2O_3–Fe_3O_4 interface, the ferrous ions react with O^{2-} or water that has diffused down through the oxide to produce mainly Fe_3O_4. Charge is built up at the Fe_2O_3–Fe_3O_4 interface rather than at the Fe–Fe_3O_4 interface. The electron deficiency of the Fe_3O_4 layer will change its chemical potential, making it harder to oxidize the iron ions and the steel samples become more resistant. The charged layer also inhibits diffusion of Fe^{2+} and O^{2-} ions.

3.11　POLYANILINE AS A BIPOLAR FILM

Wang et al. (2007) suggested that polyaniline primes were selectively permeable to anions and top coats were selectively permeable to cations. When the two coating materials are mixed together, both anions and cations can migrate through the coating. However, the polyaniline bipolar coating containing the combination of the anionic polyaniline primer with the cationic topcoat layer is a barrier for both cations and anions (Wang et al., 2007).

3.12　SYNERGISTIC EFFECT

Recently, Radhakrishnan et al. (2009) synthesized conducting polyaniline–nano-TiO_2-composite-based paint for corrosion protection of stainless steel and evaluated its corrosion resistance in 3% NaCl by electrochemical methods. The authors argued that polyaniline, being p-type, precludes electron transport, while TiO_2, being n-type, gives hindrance to hole transport across the interface. Also, the charge transport from polyaniline to TiO_2 becomes difficult because of the difference in energy levels. The authors therefore concluded that synergistic effect is responsible for exceptional corrosion protection-improvement in barrier properties, redox activity of polyaniline, large surface area available for the liberation of dopant due to nano-TiO$_2$ layer on polyaniline, and formation of p–n junction preventing charge transport in case of damaged coating. This example shows that the number of protection mechanisms may operate simultaneously as per design of the coating system.

REFERENCES

Aahmad, N., & MacDiarmid, A. G. (1996). Inhibition of corrosion of steel with the expolitation of conducting polymers. *Synth. Met., 78*, 103.

Beck, F. (1988). Electrodeposition of polymer coating. *Electrochem. Acta, 33*, 839.

Berry, D. W. (1985). Modification of the electrochemical and corrosion behaviour of stainless steels with an electroactive coating. *J. Electrochem. Soc., 132*, 1022.

Cook, A., Gabriel, A., Siew, D., & Laycock, N. (2004). Corrosion protection of low carbon steel with polyaniline: Passivation or inhibition? *Curr. Appl. Phys., 4*, 133.

Darowicki, K., & Kawula, J. (2005). Influence of polyaniline primer on the corroson properties of ST 3S steel-coating system. *Proceed of Corrosion 2005*, vol. III (pp. 197–203). Warsaw, Poland, June 8–20.

Deshpande, P. P., Vagge, S. T., Jagtap, S. P., Khairnar, R. S., & More, M. A. (2012). Conducting polyaniline based paints on low carbon steel for corrosion protection. *Prot. Met. Phys. Chem. Surf., 48*, 356.

Dung Nguyen, T., Anh Nguyen, T., Pham, M. C., Piro, B., & Takenouti, N. H. (2004). Mechanism for protection of iron by an intrinsically electronic conducting polymer. *J. Electroanal. Chem., 572*, 225.

Elkais, A. R., Gvozdenovic, M. M., Jugovic, B. Z., & Grgur, B. N. (2013). The influence of thin benzoate doped polyaniline coatings on corrosion protection of mild steel in different enviroments. *Prog. Org. Coat., 76*, 670.

Fahlman, M., Jasty, S., & Epstein, A. J. (1997). Corrosion protection of iron/steel by emeraldine base polyanaline: An X-ray photoelectron spectroscopic study. *Synth. Met., 85*, 1323.

Iroh, J., & Su, W. (2000). Corrosion performance of polypyrrole coating applied to low carbon steel by an electrochemical process. *Electrochim. Acta, 46*, 15.

Jain, F. C., Rosato, J. J., Kalonia, K. S., & Agarwala, V. S. (1986). Formation of an active elecronic barrier at Al/semiconductor interfaces: A Novel approach in corrosion prevention. *Corrossion, 42*, 701.

Kendig, M., & Warren, H. L. (2003). Smart corrosion inhibition coatings. *Prog. Org. Coat., 47*, 183.

Kinlen, P. J., Silverman, D. C., & Jeffreys, C. R. (1997). Corrosion protection using polyaniline coating formulation. *Synth. Met., 85*, 1327.

Kinlen, P. J., Menon, V., & Ding, Y. (1999). A mechanistic investigation of polyaniline corrosion protection using the scanning reference electrode technique. *J. Electrochem. Soc., 146*, 3690.

Kinlen, P. J., Ding, Y., & Silverman, D. (2002). Corrosion protection of mild steel using sulfonic and phosphonic acid doped polyanilines. *Corros. Sci., 58*, 490.

Krstajic, N. V., Grgur, B. N., Jovanovic, S. M., & Vojnovic, M. V. (1997). Corrosion protection of mild steel by polypyrrole coatings in acid sulfate solutions. *Electrochim. Acta, 42*, 1685.

Li, Y., Zhang, H., Wang, X., Li, J., & Wang, F. (2011). Role of dissolved oxygen diffusion in coating defect protection by emeraldine base. *Synth. Met., 161*, 2312.

Lu, W. K., Elsenbaumer, R. L., & Wessling, B. (1995). Corrosion protection of mild steel by coatings containing polyaniline. *Synth. Met., 71*, 2163.

Lu, W., Basak, S., & Elsenbaumer, R. (1998). Corrosion inhibition of metals by conductive polymers. In: T. A. Skothem, ed., *Handbook of Conducting Polymers*, 2nd ed. (p. 918). New York: Marcel Dekker.

Nathan, C. C. (1953). Studies on the inhibition by amines of the corrosion of iron by solutions of high acidity. *Corrosion, 9,* 199.

Ohtsuka, T. (2012). Corrosion protection of steels by conducting polymer coatings. *Int. J. Corros., 2012,* 915090, doi:1155/2012/915090.

Radhakrishnan, S., Siju, C. R., Mahanta, D., Patil, S., & Madras, G. (2009). Conducting polyaniline-nano-TiO_2 composites for smart corrosion resistant coatings. *Electrochim. Acta, 54,* 1249.

Rammelt, U., Nguyen, P. P., & Plieth, W. (2001). Corrosion protection by ultrathin films of conducting polymers. *Electrochim. Acta, 46,* 4251.

Rohwerder, M., & Michalik, A. (2007). Conducting polymers for corrosion protection: What makes the difference between failure and sucess? *Electrochim. Acta, 53,* 1300.

Santos, J. R., Mattoso, L. H. C., & Motheo, A. (1998). Investigations of corrosion protection of steel by polyaniline films. *Electrochim. Acta, 43,* 309.

Sathiyanarayanan, S., Dhawan, S. K., Trivedi, D. C., & Balakrishnan, K. (1992). Soluble conducting poly ethoxy aniline as an inhibitor for iron in HCl. *Corros. Sci., 33,* 1831.

Sathiyanarayanan, S., Muthukrishnan, S., Venkatachari, G., & Trivedi, D. C. (2005). Corrosion protection of steel by polyaniline pigmented paint coating. *Prog. Org. Coat., 53,* 301.

Sathiyanarayanan, S., Muthukrishnan, S., & Venkatachari, G. (2006). Synthesis and anticorrosion properties of polydiphenylamine blended vinyl coatings. *Synth. Met., 156,* 1208.

Schauer, T., Joos, A., Dulog, L., & Elisenbach, C. D. (1998). Protection of iron against corrosion with polyaniline primers. *Prog. Org. Coat., 33,* 20.

Tallman, D. E., Spinks, G. M., Dominis, A., & Wallace, G. G. (2002a). Electroactive conducting polymers for corrosion control: Part 2. Ferrous metals. *J. Solid State Electrochem., 6,* 85.

Tallman, D. E., Spinks, G. M., Dominis, A., & Wallace, G. G. (2002b). Electroactive conducting polymers for corrosion control: Part 1. General introduction and a review of non ferrous metals. *J. Solid State Electrochem., 6,* 76.

Wang, J., Torardi, C. C., & Duch, M. W. (2007). Polyaniline-related ion barrier anti corrosion coatings II. Protection behavoiur of polyaniline, cationic, and bipolar films. *Synth. Met. 157,* 846.

Wessling, B. (1994). Passivation of metals by coating with polyaniline: Corrosion potential shift and morphological changes. *Adv. Mater., 6,* 226.

Wessling, B. (1996). Corrosion prevention with an organic metal (polyaniline): Surface ennobling, passivation, corrosion test results. *Mater. Corros., 47,* 439.

Yan, M. C., Tallman, D. E., Rasmussen, S. C., & Bierwagen, P. (2009). Corrosion control coatings for aluminium alloys based on neutral and n-doped conjugated polymers. *J. Electrochem. Soc., 156,* 350.

4 Preparation of Protective Coatings Based on Conducting Polymers

4.1 INTRODUCTION

In early studies, the potential application of conducting polymers (CPs) in anticorrosion technology was investigated using mainly CP coatings as a primer alone or as a primer with conventional topcoats (Sitaram et al., 1997; Spinks et al., 2002; Tallman et al., 2002; Zarras et al., 2003). CP layers of a desired thickness were deposited on the metal surface either directly by utilizing electrochemical techniques such as cyclic voltammetry, potentiostatic and galvanostatic electropolymerization, or via a two-step process where a chemically synthesized CP was first dispersed/dissolved in a proper solvent and then applied, mainly by casting, on the metal. Although today, many studies still examine the protective properties of CP coating as primers to understand the factors influencing their protective performance and clarify the anticorrosion mechanism induced by CPs, several other configurations were developed to apply CPs on metal surfaces for anticorrosion purposes. Examples of such advanced configurations of CP-based coatings are multilayered CPs with suitable dopants-inhibitors and composites/nanocomposites (NCs) (Deshpande et al., 2014; Zarras & Stenger-Smith, 2014).

In general, coatings technology is a complex field, which initially evolved rather empirically. However, a deeper scientific understanding of the applicable principles was achieved within the last decades (Forsgren, 2006; Grundmeler & Simoes, 2007; Marrion, 2004; Munger & Vincent, 1999; Wicks et al., 2007). The need for a sustainable environment associated with the strict environmental regulations on the use of heavy-metal-containing paints led to seeking novel approaches to replace ingredients toxic for the human health and environment in paints. Apparently, new coatings should at least retain the protective performance of those they replace and preferably exhibit even improved properties. The investigation of CP-based coatings for the prevention of metal corrosion can be considered as being still under progress, resulting in a new technology that is developed in parallel with the general coating technology, as many well-established ingredients (binders, fillers, solvents, hardeners) utilized in traditional organic coatings are used to enhance the mechanical, processability, thermal stability, and adhesion properties of CPs.

In this chapter, CP synthesis methods and coating preparation techniques will be outlined and an assessment can be made on the basis of the obtained results in correlation with the coating performance in preventing metal corrosion. Considering the presented data, already employed techniques can be optimized or other techniques can be either devised or adopted from the literature. Section 4.2 refers to the synthesis of CPs and polymerization mechanism, while various CP formulations utilized to improve the function of CP films as protective coatings for the prevention of metal dissolution are described in Section 4.3. Selected examples of the performance of CP-based coatings are presented.

4.2 BASICS OF CP SYNTHESIS AND POLYMERIZATION MECHANISM

The synthesis of CPs and their doping have been reviewed and described in many articles and books (Chandrasekhar, 1999; Chujo, 2010; Inzelt, 2008; Skotheim & Reynolds, 2007; Wallace et al., 2003), and here, we briefly provide the basic principles and representative literature. Most CP synthesis methods are classified into two broad classes: (1) electrochemical and (2) chemical. In both classes, the synthesis of CPs involves the oxidation of the monomer to a radical cation (initiation step) and a follow-up reaction sequence in which each coupling step is activated by two species (propagation step). In this way, the polymer chain propagates until termination. In summary, the three steps, namely, chain initiation, chain propagation, and chain termination, constitute the well-known "addition or chain-growth polymerization" mechanism in polymer chemistry. All cases of electrochemical polymerization and many of chemical polymerization can be characterized as addition polymerizations. The other category is the "condensation or step-growth polymerization," which is characterized by the elimination of a chemical species when the end groups of two monomers react.

The chain initiation step in the electrochemical polymerization involves a proper imposed potential on an electrode on the surface of which the polymer will be deposited, whereas in chemical polymerization, a chemical oxidant has to be added in a batch reactor for monomeric radical cations to be formed for the initiation of the polymerization process. There is another difference between electrochemical and chemical polymerization regarding the mechanism and, in particular, the chain propagation step. In the electrochemical polymerization, a radical–radical coupling is suggested to occur owing to the abundance of radical cations adjacent to the electrode surface where the reaction occurs, while in the chemical polymerization, the radical cation attacks a monomer molecule. Some details on the mechanism and implementation of electrochemical and chemical polymerizations are given further in the chapter.

4.2.1 Electrochemical Oxidative Polymerization

Electrochemical polymerization is considered as the most important method for preparing CP homogeneous films of a desired thickness and morphology, adherent to a metal or semiconductive substrate electrode. The electrochemical polymerization process has attracted special attention over the last decade for the fabrication of several nanostructures of CPs (Cho & Lee, 2008; Guo & Zhou, 2007). Electrochemical polymerization provides various advantages as compared with the chemical polymerization: (1) enhanced adhesion to the metal substrate, (2) rapid polymerization at room temperature, (3) simultaneous doping with the desired doping anion, which is often chosen to be the anion of the supporting electrolyte in the polymerization mixture,

(4) control of the film thickness and perhaps molecular structure and morphology, and (5) avoidance of solubility problems encountered with several chemically synthesized CPs to be applied on metal surfaces. The drawbacks of electrochemical polymerization arise first when polymerization on oxidizable metals is necessary to occur from acidic solutions (e.g., deposition of polyaniline), as metal electrodissolution may occur and, second, when electrodeposition on large surfaces, such as bridges, ships, and pipelines, is required. Then, electrochemical polymerization is indeed cumbersome and practically impossible. In these cases, chemical deposition in conjunction with other techniques such as casting and blending is preferable.

The addition mechanism for CP synthesis from a monomer, RH_2, can be described in terms of electrochemical reaction mechanisms as $E(C_2E)_nC$, and can be analyzed in terms of the following steps:

$$RH_2 \rightarrow RH_2^{+} + e \qquad (E)$$
$$2RH_2^{+} \rightarrow [H_2R\text{-}RH_2]^{2+} \rightarrow HR\text{-}RH + 2H^+ \quad (CC)$$
$$HR\text{-}RH \rightarrow [HR\text{-}RH]^{+} + e \qquad (E)$$
$$[HR\text{-}RH]^{+} + RH_2^{+} \rightarrow [HR\text{-}RH\text{-}RH_2]^{2+} \rightarrow HR\text{-}R\text{-}RH + 2H^+ \rightarrow\rightarrow\rightarrow$$
$$(HR\text{-}R\text{-}RH)_n + H_2O \rightarrow \text{termination} \qquad (C)$$

Figure 4.1 illustrates an example of the electrochemical polymerization of a heteroaromatic monomer, where X might be NH (pyrrole, Py), S (thiophene, Th), or O (furan), leading to polypyrrole (PPy), polythiophene (PTh), and polyfuran, respectively. Polyfuran and its derivatives have not been investigated with respect to metal corrosion protection rather because of their relatively high oxidation potential that results in a material with poor processability and low conductivity. Moreover, it seems that polyfuran cannot be reoxidized by the oxygen reduction reaction, a necessary process for the active role of CPs in anticorrosion control. Although Th exhibits also a high oxidation potential, for instance, ~1.8 V_{SCE} in the $CH_3CN\text{-}LiClO_4$ solution, electrochemical polymerization on oxidizable metals has been carried

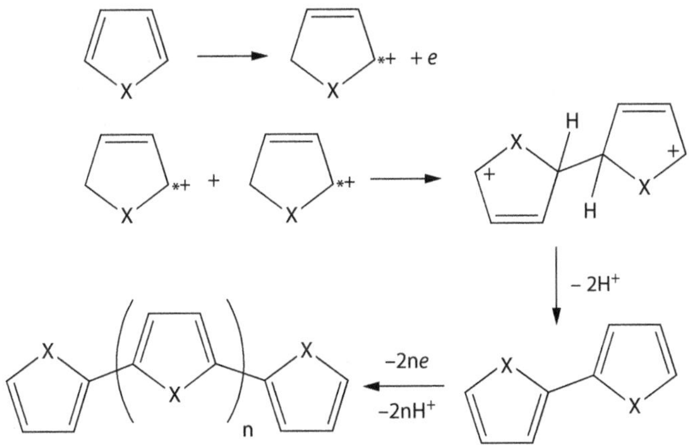

FIGURE 4.1 Schematic electrochemical polymerization of heteroaromatic monomers, where X = NH (pyrrole) or S (thiophene).

out starting from bis-thiophene, whose oxidation potential is essentially lower, at ~0.7 V_{SCE}, in mixed aqueous-organic media than that of Th (Barsch & Beck, 1993).

As can be seen in Figure 4.1, the polymer initiation step occurs by the anodic oxidation of a suitable monomer to its radical cation upon imposing a proper potential depending on the standard oxidation potential E^0 of the monomers. The oxidation potentials of several monomers giving rise to CPs, which might be of interest in metal corrosion control are summarized in Table 4.1. The fate of monomer radical cations formed in the vicinity of the electrode depends on their stability. Most frequently, the monomer radical cations undergo a radical–radical coupling leading to dimerization and, in turn, to polymerization. However, if the radical cations are very stable, they may diffuse away from the electrode and form soluble (low-molecular-weight) products reacting with neutral monomers or other soluble products; on the other hand, if they are very unstable, they may react randomly with the supporting electrolyte anions or the solvent, leading again to soluble products. A successful electropolymerization leading to the growth of a polymer with desired properties is often a matter of choosing the appropriate conditions, such as electrical parameters, deposition substrate, solvent, and supporting electrolyte.

The electrochemical polymerization of Py or Th results in the PPy or PPh in which the monomers are coupled through the 2-, 5-positions (Waltman & Bargon, 1986). In general, the "ideal" structure of polymers can be roughly predicted from the symmetry and reactivity of the monomer radical cation. Depending on the unpaired electron distribution in the oligomeric radical cations, different linkage and "nonideal" structures may be also obtained. However, considering that oligomeric radical cations are less reactive than monomeric ones, it seems that the monomer radical cation is the critical factor of linkage sites. The reactive sites on monomers and the ideal polymer structure expected for most of the CPs investigated for their anticorrosion properties can be seen in Table 4.1.

The mechanism of aniline (AN) electrochemical polymerization has been discussed extensively in literature. In particular, the initiation of polymerization is strongly dependent on the applied potential and pH of the polymerization solution. Both factors affect not only the oxidazibility of AN and its protonated form (anilinium cation, $pK^a = 4.6$) but also the redox/acid-base properties of its reactive species formed by single-electron (anilinium dication radical and aniline cation radical/neutral radical) and two-electron (aniline dication/nitrenium cation/nitrene) oxidation (Ciric-Marjanovic, 2013a). The mechanism of dimerization reactions and, in turn, of the polymerization of AN depends on the pH of the polymerization solution. The film adsorbed at the electrode is found, by Fourier transform infrared (FTIR) spectroscopic characterization, to be a "head-to-tail" (N-C4) polymer, whereas "tail-to-tail" (C4–C4) free radical recombinations of AN cation radicals (benzidine) remain in the solution. Formation of benzidine is facilitated in more acidic solutions.

The electrochemical growth of polyaniline (PAN) films was suggested to occur upon the basis of p-aminodiphenylamine aniline dimer. In acidic solutions, it is characterized as autocatalytic in the sense that the highest oxidized PAN (pernigraniline) attacked by an AN molecule grows to a longer chain, and this explains why, after the monomer oxidation at 0.9 V_{SCE} (V vs. SCE, saturated calomel electrode), the film may continue to grow even at lower potentials where the monomer oxidation is not predicted to occur

Table 4.1
Oxidation Potentials of Monomers and Linkage Sites (Indicated by Arrows) Considered with the Highest π-Electron Density in Monomer Radical Cations for the Electrochemical Preparation of CPs Utilized in Coatings for Metal Protection

Monomer	Monomer Structure	Electrophilic Reactive Sites	E_{ox} (V_{SCE})[a]	Ideal Polymer Structure
Pyrrole			0.8	
Thiophene			1.6	
Aniline			0.9	
Indole			0.9	
Carbazole			1.2	

[a] The oxidation potentials represent indicative values for the electrochemical oxidation of heteroaromatics in CH_3CN and of AN in acidic solutions.

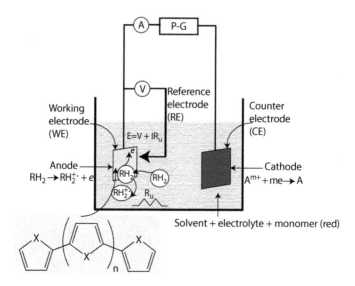

FIGURE 4.2 Schematic illustration of a three-electrode electrochemical cell connected with a potentiostat-galvanostat (P-G) utilized for the electrodeposition of CPs.

(Gospodinova & Terlemezyan, 1998; Stilwell & Park, 1988). The oxidation potential of oligomers in a pernigraniline form increases with its protonation degree, facilitating the formation of PAN with a higher molecular weight in strong acidic media. The acidity of the solution determines also the linearity of the oligomer/polymer of AN. The linear/branched ratio increases with decreasing pH (Ohsaka et al., 1984).

Electrochemical polymerization can be implemented by means of cyclic voltammetry (cyclic potential sweep, CPS) deposition, potentiostatically or galvanostatically. A typical three-electrode cell used in electrochemical polymerization is shown in Figure 4.2. The three electrodes, namely, the working electrode (WE), reference electrode (RE), and auxiliary or counter (CE) electrodes, are combined appropriately with a potentiostat/galvanostat (details can be seen in Bard & Faulkner, 2000). A two-electrode cell consistent with the WE and CE electrodes can be also employed.

4.2.1.1   CPS Deposition

In this case, the potential is swept between two values, E_1 and E_2, at a fixed potential sweep rate, and when the potential reaches value E_2, the sweep is reversed and the potential is swept back to E_1. Thus, a triangle potential pulse is applied repeatedly (Figure 4.3a) and the current response is recorded vs. the applied potential E, called cyclic voltammogram (Figure 4.3b). The criteria for the selection of the scanned potential region are based on the oxidation potential E^0 of both the monomer and the metal that acts as the WE; the monomer should be oxidized within this region, while the electrodissolution of the metal-substrate should be avoided in the utilized electrolytic medium, if possible. As in the investigation of every electrochemical process, the electrolyte and solvent should be inert within the scanned potential region.

Figure 4.3 shows an example of cyclic voltammograms obtained during AN oxidation on stainless steel (SS) of AISI 304 (first potential cycle within the potential range

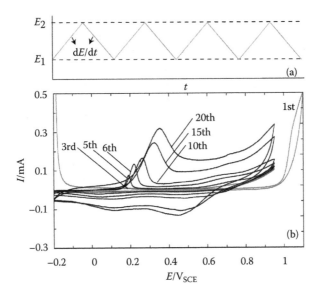

FIGURE 4.3 (a) Repeated triangle potential pulses applied at SS where deposition of AN is expected to occur. (b) Cyclic voltammograms traced at a sweep rate $dE/dt = 20$ mV s^{-1} during the CPS deposition of PAN from 0.5 M H_2SO_4 containing 0.1 M AN. Oxidation of AN ($E^0 \sim 1$ V) on SS (AISI304) occurs during the first potential cycle scanned in the range between -0.2 and 1.1 V. Subsequent cycles scanned in the range -0.2 to 0.95 V indicate the redox processes of oligomer/polymer occurring at potentials (0–0.6 V) lower than E^0.

-0.2 to 1.1 V_{SCE}) and the redox processes of the oligomers/polymer taking place at $E < E^0$ during the subsequent potential cycles within the potential range -0.2 to 0.95 V. The high current values observed at -0.2 V is a result of the electrodissolution of SS. The electrodissolution of SS is suppressed during the second and subsequent cycles because of the oxide formation and surface passivation of the SS substrate. Therefore, deposition of the PAN film occurs on a passive surface, resulting in cyclic voltammograms similar with those observed during electrodeposition of AN on a noble metal (i.e., Pt, Au). CPS deposition of the PAN becomes evident by the gradual increase in the current density observed at $E < E^0$ during successive potential cycling, indicating the formation of AN oligomers and, in turn, polymer growth. The PAN growth occurs during successive potential cycling within the range -0.2 to 0.95 V (Figure 4.3b). A lower potential limit E_2 is used to avoid overoxidation of PAN. It was observed that despite that $E_2 < E_0$, PAN deposition occurs since the polymerization of AN follows an autocatalytic mechanism (Stilwell & Park, 1988). In the case of CPS deposition, AN cation radicals formed during the first potential cycle initiate polymerization that in subsequent cycles proceeds even at lower potentials as oxidized oligomer/polymer acts as oxidant incorporating oxidized monomers in the polymer chain.

4.2.1.2   Potentiostatic Deposition

The excitation in chronoamperometric experiments for CP deposition is a potential step applied to the working electrode, and once the double layer is charged, the potential remains constant (Figure 4.4a). The measured response of the ensuing electrode reaction is a current–time (I–t) curve (Bard & Faulkner, 2000), with a decaying current reaching a steady-state value at a certain time. In the case of an electrochemical polymerization on an oxidizable metal (e.g., SS, Fe), the initial decaying current, owing to the electrodissolution and then to passivation of the metal substrate, rises with time as the deposition of the electroactive CP film proceeds. At a later time, the current decreases because of decreasing conductivity of the deposited film with an increase in its thickness. An example of the applied excitation potential and I–t response obtained during AN polymerization is illustrated in Figure 4.4b for the AN potentiostatic polymerization from 0.5 M H_2SO_4 containing 0.1 M AN. As can be seen, the shape of the I–t curves depends on the excitation pulse and

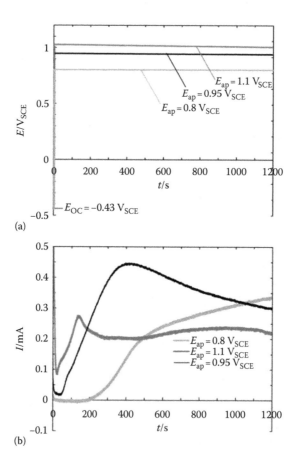

FIGURE 4.4 **(See color insert.)** (a) Potential step (E–t) applied to the SS (AISI 304) electrode between the open circuit potential, E_{OC}, of the SS in 0.5 M H_2SO_4 and a constant potential, E_{ap}, chosen to be suitable for PAN deposition, and (b) current–time (I–t) response of the AN electrochemical polymerization in 0.5 M H_2SO_4 containing 0.1 M AN. Depending on the E_{ap}, the basic current is different because of SS reactions, namely, either passivation or transpassive dissolution beginning at $E \sim 1$ V_{SCE}.

applied potential, E_{ap}, which determine the occurrence of electrode reactions, including the oxidation of the metal substrate itself, the monomer oxidation, oligomerization, polymerization, and the CP film electroactivity. Chronoamperometric experiments for the CP deposition are not restricted to a simple potential pulse. Multiple potential pulses might be applied, consisting of a more positive potential, causing oxidation of the monomer, followed by a less positive potential, causing reduction in CP.

4.2.1.3 Galvanostatic Deposition

The galvanostatic method is the inverse to the potentiostatic one in the sense that a constant current pulse is applied to the working electrode (Figure 4.5a). Then the potential shifts from the equilibrium and changes are recorded as a function of time, resulting in a potential–time (E–t) curve (Figure 4.5b). For example, in the case of the PAN deposition on SS (Figure 4.5b), the constant current chronopotentiometry involves a certain transition time during which the SS electrodissolution occurs at low potentials (~0.4 V_{SCE}), followed by the SS passivation recognized by the potential transition to relatively high values (~1 V_{SCE}). The initial transition period decreases by increasing the applied current, I_{ap}. At the higher potential, the monomer oxidation

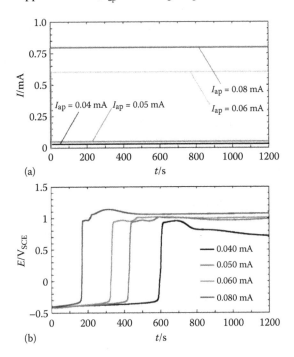

(a)

(b)

FIGURE 4.5 **(See color insert.)** (a) Current step (I–t) applied to the SS (AISI 304) electrode being at the open circuit or equilibrium potential of the SS in 0.5 M H_2SO_4, where $I = 0$ to a current I_{ap}, and (b) potential–time (E–t) response including the electrodissolution and passivation of SS followed by the AN oxidation and its subsequent polymerization in 0.5 M H_2SO_4 containing 0.1 M AN. Depending on the I_{ap}, the time required for SS passivation differs, whereas different potentials are also established for the polymerization process.

is feasible along with the initiation of the polymerization reaction on the passive SS surface. The potential at which the polymerization process occurs depends strongly on I_{ap}, resulting in CP films of different properties. PAN films of different morphologies and molecular structures have been received.

Electrochemical techniques are employed not only for the formation of homopolymer films. Electrochemical copolymerization leading to the formation of CP copolymers or multilayers of the same or different CPs of a particular micro/nanostructure on oxidizable metal substrates can also be carried out, readily providing well-designed adherent coatings. The electropolymerization approach offers advantages over other coating technologies, the main advantage being the ability to form the polymer at irregular shapes of different micro/nanostructure and or combine different electrochemical techniques to control the rate of polymerization stages. Therefore, in recent years, there is much interest in developing innovative electropolymerization procedures that give rise to homogeneous and adherent CPs with large active surface trapping different organic and inorganic fillers.

4.2.2 CHEMICAL OXIDATIVE POLYMERIZATION

Chemical oxidative polymerization is used extensively in preparing CP-based coatings because it is a simple way for large-scale production of CPs, required for instance in the case of protecting large metal surfaces and where application of multilayers of CPs and non-CPs is required. Drawbacks of chemical polymerization in the case of using CP as primers are (1) the limited solubility of CPs, and especially of PAN, in most of the common organic solvents and (2) the poor adherence between CPs and metal substrate. For example, PAN, the most widely used CP in many applications, is insoluble in almost all common solvents, and several approaches are frequently used to improve PAN solubility and processability. These approaches include emulsion polymerization and synthesis of colloidal PAN by doping with polymeric stabilizers or electrostatic stabilizers that are suitable doping anions, inserted either by using anionic surfactants or functionalization of the monomer or polymer (Wallace et al., 2003). The presence of immobilized dopants, such as sulfonic groups, in the polymer chain increases the solubility in polar solvents (Freund & Deore, 2007). For the enhancement of the adhesion of PAN-based coating on metal surfaces, appropriate binders are frequently used.

In the case of chemical oxidative polymerization, an appropriate oxidant initiator replaces the imposed potential required in electrochemical polymerization. A variety of such oxidants used for CP chemical synthesis with their standard oxidation potential, E^0, can be seen in Table 4.2 (Wallace et al., 2003).

Under the term *chemical oxidative polymerization*, there is a variety of synthetic routes, such as emulsion polymerization, enzyme catalytic polymerization, and interfacial polymerization, used extensively in designing CP morphologies and properties required for specific applications.

Although the majority of investigations on CP-based protective coatings are dominated by the preparation of formulations with main ingredients PAN, PPy, PTh, and their derivatives, other CPs were also used in metal corrosion protection, such as poly(para-phenylene vinylene) (PPV) derivatives via a precursor method (Wessling,

Table 4.2

Oxidants and Their Standard Oxidation Potential, E^0, Utilized for CP Synthesis Through Chemical Polymerization

Oxidant	E^0 $(V_{SCE})^a$
$(NH_4)_2S_2O_8$	1.698
H_2O_2	1.538
$Ce(SO_4)_2$	1.478
$K_2Cr_2O_7$	0.988
KIO_3	0.848
$FeCl_3$	0.528

1985), which differs from those of the previously mentioned CPs. It involves the synthesis of the bis-sulfonium salt of 1,4-bis(chloromethyl)benzene, followed by sodium hydroxide elimination and polymerization at a low temperature to give an aqueous solution of a precursor polymer. This soluble precursor polymer can be processed and converted into PPV by thermal elimination. The improvement of the processability and stability of precursor compounds through synthetic methods led to functionalization of the PPV and the poly(2,5-bis(*N*-methyl-*N*-hexylamino)phenylene vinylene (Irvin et al., 2002), used as protective coatings on Al alloys (Anderson et al., 2002; Stenger-Smith et al., 2004).

4.2.3 SYNTHESIS OF FREQUENTLY USED CPs IN ANTICORROSION PROTECTION

4.2.3.1 Synthesis of PAN

Electrochemical polymerization of AN is a versatile route for PAN synthesis resulting in PAN films of different electroactivities and a variety of other properties depending on the anode potential. Suitably choosing the anode potential and pH of the polymerization solution would avoid overoxidation and degradation. The electrochemically promoted degradation of PAN in highly acidic solutions leads to the 1,4-bezoquinone as the major product, although other species have been also identified. It is suggested that the presence of hydroquinone/1,4-benzoquinone enhances the stability of PAN structure, namely, increases its resistance to overoxidation, preventing further degradation to a certain extent (Ciric-Marjanovic, 2013a).

Electropolymerization of AN has been carried out in many active metals and alloys (Biallozor & Kupniewska, 2005), such as Fe (Bernard et al., 2001; Camalet et al., 1996; Sazou, 2001; Sazou & Georgolios, 1997), Zn (Lacroix et al., 2000), Ni (Prasad & Munichandraiah, 2001), and SS (Kraljic et al., 2003; Obaid et al., 2014; Ozyilmaz et al., 2006b; Qin et al., 2010; Sazou et al., 2007) from both aqueous and nonaqueous media. Electropolymerization of AN on SS, Ni, Ti, Al, and Pb has been studied in aqueous solutions of Na_2ClO_4, $H_2C_2O_4$, and H_2SO_4 (Prasad & Munichandraiah, 2001). The electroactivity of the resulting PAN films was investigated by the cyclic voltammetry method in comparison with PAN electrodeposited on

Pt. It was found that the rate of Fe^{2+}/Fe^{3+} redox reaction is greater on PAN-modified Ni than Pt electrode. Preparation of PAN coating on magnesium alloys by pulse potentiostatic methods has been performed (Guo et al., 2003). The characteristics of PAN coating prepared by various methods, namely, galvanostatic, potentiostatic, and potentiodynamic, have been given recently (Mondal et al., 2005). The electrochemical properties of the PAN film in dependence of the electropolymerization way were examined as well (Sazou et al., 2007).

The chemical synthesis of PAN has been known since the 19th century as "aniline black" and utilized in textiles as dyes and in printing. Its polymer nature was not known at that time. As the most utilized CP in many applications, and in protective coatings technology, numerous chemical approaches for PAN synthesis have been reported in the literature (Ciric-Marjanovic, 2013a, and references therein). The most frequently utilized oxidants are ammonium persulfate (APS) and Fe(III) compounds (i.e., $FeCl_3$). Moreover, H_2O_2, KIO_3, as well as transition metal compounds (Mn, Cr, Ce, V, Cu), noble metal compounds (Au, Pt, Pd, Ag), mixtures of oxidants ($FeCl_3$/H_2O_2, KIO_3/NaClO), and addition of various enzymes were also used. An example of the chemical polymerization of AN by APS is shown in Figure 4.6. Using NH_4OH ensures that the obtained PAN is in its emeraldine base (EB) state, allowing insertion of a desired dopant at a later stage.

The incorporation of ionogenic acid groups (i.e., sulfonate, carboxylate) into the AN ring on 2-, 3-, 4-positions and N-atom, either before or after polymerization, has let to self-doped, water-soluble PAN (Freund & Deore, 2007). Besides solubility and

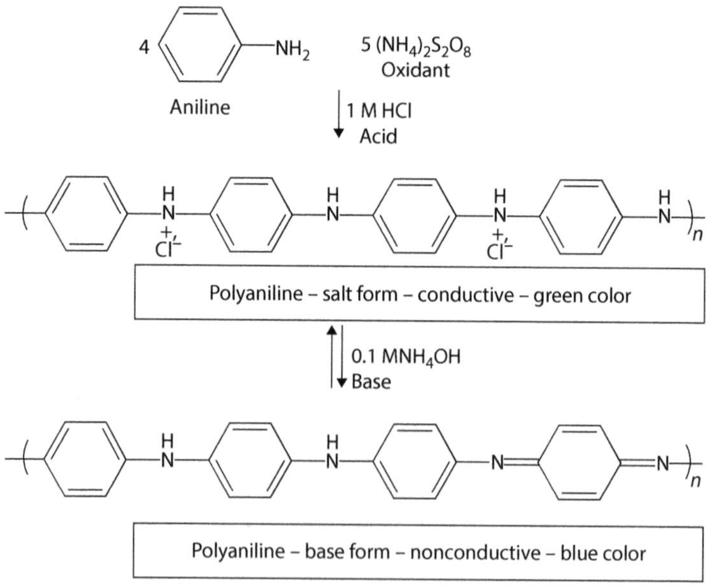

FIGURE 4.6   Chemical oxidative polymerization of AN using APS as oxidant and dedoping process by a treatment with NH_4OH to obtain EB. (From Ciric-Marjanovic, G., *Synth. Met.*, *177*, 1–47, 2013.)

processability, self-doping affects also other properties of PAN, like redox behavior and morphology (Michael et al., 2015). In the case of PAN-based protective coatings, self-doping may prevent ionic cross-exchange, for example, in halide-containing corrosive media.

4.2.3.2 Synthesis of PPy

Electrochemical polymerization is a versatile method used very often for the preparation of PPy and its doping. This synthetic route leads to high-quality films of high conductivity (up to 10^2 s cm^{-1}) (Biallozor & Kupniewska, 2005). First attempts to deposit PPy on oxidizable metals (Fe, Cu, Ti, SS) were reported by Schirmeisen and Beck (1989) using several aqueous and nonaqueous electrolytes. Dissolution of Cu and Fe was observed instead of film deposition in most cases, whereas successful formation of PPy film was achieved on Fe from nitrate salts. Later on, oxalic acid was shown to be a suitable electrolyte for PPy electrodeposition on oxidizable metals, similarly with PAN mentioned previously. Strongly adherent smooth PPy layers were formed by galvanostatic (current-controlled) deposition on Fe because of the formation of a passivating iron (II) oxalate interlayer (Beck & Michaelis, 1992; Beck et al., 1994; Ferreira et al., 1996). At constant current, it was found that pretreatment of mild steel and iron surfaces by 10% nitric acid inhibits iron dissolution and allows electroplymerization of Py in Na_2SO_4, $K_2C_2O_4$, and KNO_3 with a maximum coulombic efficiency of 95% in Na_2SO_4 for a current density of <10 mA cm^{-2} (Ferreira et al., 1996).

Successful deposition of PPy from oxalic acid solutions was also carried out on Al surfaces (Hulser & Beck, 1990). In the latter case, pretreatment of the metal by polishing or by anodic galvanostatic activation was an essential step. In all cases, the Al_2O_3 surface layer with pores, usually filled with electrolyte, was transformed to a Al_2O_3 layer with PPy filled pores.

Different types of an electrochemical pretreatment of the substrate surface were devised for a successful electropolymerization of PPy on Zn (Petitjean et al., 1999, 2005), such as the following:

1. Electrochemical pretreatment in 0.2 M Na_2S, which led to the formation of mixed layers containing ZnS and ZnO_xH_x, whose composition appears to be constant for a certain current intensity regardless of the Na_2S concentration. It was found that the inhibiting effect of sulfides is more effective than that of hydroxides, resulting in an essential decrease in the Zn dissolution current and passivation of the Zn surface. Galvanostatic electrodeposition of PPy can be then carried out from oxalic acid solutions of pH 6 at certain current values (Zaid et al., 1998).
2. Electrochemical pretreatment with hexacyanoferrate preceding either a two- or one-step electropolymerization of Py (Pournaghi-Azar & Nahalparvari, 2005). The two-step process includes (1) pretreatment of the substrate surface by the formation of zinc hexacyanoferrate (ZnHCF) as an effective protecting layer to prevent the Zn anodic dissolution and (2) electropolymerization of Py on the Zn electrode covered with the ZnHCF film.

In the one-step process, both the pretreatment of the Zn surface and Py electropolymerization are achieved simultaneously in the solution containing $Fe(CN)_6^{4-}$ and Py. Both the one-step and two-step approaches led to the electrochemical oxidation of the Zn and the formation of potassium ferricyanide with the metal ion, which appeared as a thin protective layer that is irreversibly attached to the electrode matrix via a CN ligand-bridged dinuclear Zn–Fe species, allowing the formation of adherent and homogeneous PPy films. With the one-step approach, the deposition of PPy coatings on Zn is very fast and does not require any pretreatment of the metal surface. The PPy films obtained by both approaches on Zn are very adherent and can be utilized as primer layers for the protection of Zn in corrosive environments.

The chemical synthesis of PPy was first reported in 1916 via the oxidation of Py by using H_2O_2 as oxidant, and the product was called as "pyrrole black" (Myers, 1986). Nowadays, $FeCl_3$ is frequently used as oxidant to produce PPy. This synthetic route is simple but may lead to insoluble and unprocessable PPy and hence less suitable for preparing anticorrosion coatings. Regardless of the polymerization route, to render PPy more soluble in organic or aqueous media, alkyl or alkylsulfonate groups have been attached to the monomer before the polymerization (Patil et al., 1987). Electropolymerization of alkylated Py led to solubilities of about 400 g L^{-1} in organic solvents and lower, but still good, conductivities (1–30 S cm^{-1}) (Ashraf et al., 1996). Both chemical and electrochemical polymerizations of 3-alkylosulfonated pyrroles led to polymers of a lower conductivity (10^{-3}–0.5 S cm^{-1}) but of a higher solubility in organic solvents and, in particular, in aqueous media (Havinga et al., 1989). Alkylosulfonate derivatives via the N-position were also electrochemically polymerized to produce PPy derivatives of enhanced solubility. Because of the even lower conductivity of this homopolymer, copolymerization with Py was suggested to enhance polymer conductivity (Reynolds et al., 1988). Although Py chemical polymerization is an easy synthetic route, electrochemical polymerization is a more versatile route for the deposition of PPy on active metal surfaces.

4.2.3.3 Synthesis of PTh

Electrochemical polymerization to produce PTh from Th monomer or its oligomers, which have a lower oxidation potential, mostly in organic media such as acetonitrile or propylene carbonate, is generally a simple route (Aeiyach et al., 1997). The main reasons that preclude the electropolymerization of Th in aqueous media are (1) its low solubility, (2) the high reactivity of its radical cation with nucleophilic media (i.e., water), and (3) its high oxidation potential. Several approaches have been proposed to achieve Th electropolymerization in aqueous media, including the use of water–thiophene emulsion, 85% H_3PO_4 for the polymerization of Th oligomers, previously produced by chemical oxidation of Th in acid solution, a neutral micellar medium in water by using sodium dodecyl sulfate (SDS) as the surfactant mixed with a small amount of butanol in aqueous $LiClO_4$ (Bazzaouri et al., 1996). This latter approach gave inhomogeneous but adherent films on both Pt and Fe substrates using bis-thiophene that was oxidized at much lower potentials ($0.7\ V_{SCE}$) as compared with Th (Barsch & Beck, 1993).

The chemical polymerization of Th via oxidative coupling leads to PTh, which is, same as PPy, also insoluble and unprocessable. However, PTh can be rendered soluble in either organic solvents or aqueous media using similar functionalization approaches used in the case of PPy. Incorporation of functional groups, such as alkyl groups in the 3-position, led to PTh derivatives (poly(3-alkylthiophenes) (Loewe et al., 2001) with sufficient conductivity as well as improved processability and solubility in organic solvents for utilizing them as anticorrosion coatings of metals.

4.2.4 SYNTHESIS OF NANOSTRUCTURED CPs

Intensive research on nanostructured CPs during the last years has led to a new perspective for many applications. Nanostructures of CPs are currently of great interest since they combine the properties of low-dimensional organic conductors with high-surface-area materials and, frequently, remarkably enhanced processability, all these properties being required in corrosion protection (Tian et al., 2014; Yang et al., 2010; Yao et al., 2008). New synthetic and characterization methods have been applied to designing nanomaterials with specific physicochemical properties (Eftekhari, 2010).

Nanostructures of CPs can be prepared chemically or electrochemically using a variety of templates or template-free procedures. Hard and soft templates can be used for the synthesis of CP nanotubes and nanofibers such as PPys, PThs, and PANs in the pores of a polycarbonate or alumina membrane (Xiao et al., 2007). Surfactants and/or suitable dopants acting like surfactants constitute approaches attracting great interest to synthesize CP nanostructures (Lee et al., 2008; Li et al., 2008; Liao et al., 2012; Zhou et al., 2009). For example, self-assembly process conducted to prepare self-doped sulfonated polyanilines (SPANs) can lead to a variety of morphologies (Lee et al., 2008). The morphology of SPANs can be changed from microspheres and nanotube to coral-reef-like structures by simply increasing the molar ratio of the monomer (AN) to the o-aminobenzesulfonic acid, acting as the dopant anion and surfactant.

Nanofibers of biodegradable polymer can be an alternative choice as a template: The conductive polymer is electrodeposited on the surface of electrospun nanofibers, which are removed to generate hollow conductive polymer nanotubes. In these studies, the growth of nanotubes on a template can be explained by the mechanism based on the interaction, such as solvophobic and electrostatic, between conductive polymer and a template.

Electrochemical polymerization is an environmentally friendly method used to produce tunable CP nanostructures, in particular, PAN nanotubular/-globular structures with tunable size and shape, which are great complements to chemical methods. Electrochemical methods do not need any surfactant and oxidant and allow control of the rate of polymerization reaction by varying the imposed anodic potential/current. Several factors such as dopant, current density, and potential have great effects on the morphology of CP obtained by electrochemical polymerization (Guo & Zhou, 2007).

4.3 CP-BASED PROTECTIVE COATINGS

The protective performance of CPs was examined utilizing them as primers with or without conventional top coats, as blends with a conventional polymer coating, as additives to modify a conventional organic coating, and as composites/NCs with inorganic materials or with carbon black and carbon nanotubes. Moreover, bi/ multilayers consisting of combination of different CPs, like PAN and PPy (Tan & Blackwood, 2003) and copolymers of CPs (Çakmakcı et al., 2013), have been also utilized, especially in cases of poor adhesion of the homopolymer layer on metal surfaces, to improve the adherence of CP coating or to reduce coating permeability against water ingress (Yalcinkaya et al., 2010).

In the scheme depicted in Figure 4.7, most of the approaches that investigated developing CPs as protective coatings are summarized. This classification is somehow an arbitrary one and we do not claim to provide a complete summary in each case but aim that many of the key procedures are included. As can be deduced from previous chapters, the choice of a CP formulation depends on a vast variety of factors including the structural material (chemical nature, size, and shape), the period for which protection is required, and the composition and pH of the corrosive environment.

Recent advances in corrosion protective CP-based coatings focus on organic/ inorganic composite coatings. They offer reliable long-term protection to the metal substrate by exploiting the barrier effect and self-healing properties provided by the organic part of the coating in conjunction with the inhibiting effect provided by inorganic pigments/inhibitors.

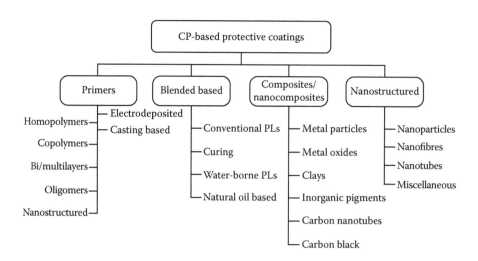

FIGURE 4.7 Various forms of CP-based protective coatings.

4.3.1 CPs as Primers

CPs have been used as primers directly deposited onto the metal as a film by means of electrochemical and chemical procedures. The first studies devoted to the protective properties of CP films and, in particular, of PAN films examined the functionality of CP as a primer against metal corrosion. Those studies on CP primers contributed significantly in understanding several aspects of the protection mechanism offered by CPs to metals, especially to iron and steels. Understanding the corrosion protection mechanism requires consideration of the charge transfer reactions occurring across the metal/CP and CP/electrolyte interfaces, along with the control of permeability and selectivity of polymers to ionic exchange processes occurring across the CP/electrolyte interface. An unambiguous conclusion from first studies, for example, in the case of PAN deposited on Fe, was the enhancement of Fe passivation, which manifested itself by a significant shift of the E_{OC} to higher values located within the passive potential region and a simultaneous increase in the polarization resistance, R_p. An effective corrosion protection was also observed by PAN for other metals, such as Al and Cu. Similar positive results were gradually arisen regarding PPy deposited onto Fe, steels, and other metals. However, it was recognized early that these promising signs regarding the protective abilities of CP primers last within a relatively short time as compared with conventional organic coatings because of blistering and delamination of the CP primer. Thus, top-coated CP primers were suggested instead of simple CP films by exploiting the barrier properties of epoxy top coats of sufficient water permeability (Shauer et al., 1998). A top coat is crucial for the durability of an active corrosion protection of the CP coating.

4.3.1.1 Electrodeposited CP Coatings

Electrodeposition of CPs from a solution of an electrolyte and monomer in a conventional three-electrode electrochemical cell connected with a potentiostat-galvanostat (Figure 4.2) is a versatile technique to form CPs of different micro/nanostructures as well as CP-based composites and NCs. The factors affecting the properties of the deposited coating are first closely related with the choice of the electrochemical technique, the electrical parameters used during electroplymerization (i.e., potential region, potential sweep rate, applied current or applied potential), and the parameters influencing the electrochemical processes (solvent, electrolyte, pH, temperature, etc.) (Breslin et al., 2005; Lacaze et al., 2010; Santos et al., 1998; Sazou & Georgolios, 1997; Sazou et al., 2007).

The electrolyte, which provides the polymer dopant, is of critical importance for the performance and adherence of the CP coating. The polymerization solution in the case of PAN deposition should be mostly acidic (pH <4) to obtain an electroactive PAN film. A problem that had to be overcome in early stages of the development of the application of CPs in anticorrosion technology was the increased dissolution of metal substrate in acidic solutions during electrodeposition. In the case of steels, this problem was addressed by using mainly oxalic acid solutions because oxalates react with ferrous ions during an initial electrodissolution step, resulting in the formation of a thick layer of ferrous oxalate (Bernard et al., 1999; Camalet et al., 1996, 1998a; Mrad et al., 2009; Sazou & Georgolios, 1997).

Polyaniline: Since the first report by DeBerry (1985) on AN direct polymerization on SS (AISI 410 and 430) from sulphuric acid solutions, a large number of studies have reported direct deposition of PAN films on active metals. For instance, electrodeposition of PAN thin films was carried out directly on Fe and steels from oxalic acid solutions (Bernard et al., 1999; Camalet et al., 1998a; Sazou & Georgolios, 1997). A passivating ferrous oxalate layer reduces remarkably the Fe electrodissolution, thus facilitating PAN growth. Although the positive shift of open circuit potential, E_{OC}, in sulfuric acid solutions readily shifted to negative values close to E_{OC} values corresponding to the Fe–H_2SO_4 interface, lower dissolution currents were obtained. The E_{OC} remained in the passive state for relatively longer periods (~10 h) in less acidic solutions (i.e., sulfate medium of pH 4.4). Other polymerization electrolytes were also used, such as phosphoric acid, for which particular optimized conditions are suggested for the potentiostatic growth of PAN on Fe (Bernard et al., 2001). A comparison of the protective properties of PAN layers deposited on steel samples (13% and 4.44% Cr) from phosphate and sulfate solutions shows that PAN films from phosphates appear to have better protective properties than the layer deposited in a sulfate solution. However, in the chloride-containing solution, the time of protection was significantly shorter (Kraljic et al., 2003).

In the case of SS, the sulfuric acid solution seems to be a proper polymerization medium, as electrodissolution of SS (AISI 304) occurs only during the first potential cycle (Figure 4.3). In fact, polymerization rates of AN on SS from sulfuric acid solutions seem to be comparable or higher than those obtained for polymerizations on "inert" metals (Pt, Au, glassy carbon [GC]) (Ozyilmaz et al., 2004, 2006a,b; Sazou et al., 2007). Cyclic voltammetric and potentiostatic polymerization leads to the growth of PAN on the passive SS, resulting in long-term protection in chloride-free sulfuric acid solutions as a result of the formation and stabilization of a dense oxide film on the SS surface. However, in chloride-containing sulfuric acid media, the durability of protection is limited to hours because of the insertion of Cl⁻ and initiation of localized corrosion.

Dense oxide film due to the presence of PAN was also responsible for the improved corrosion protection of PAN coating on the AISI 430 SS substrate. PAN was electrodeposited on the surface of AISI 430 SS in nitric acid solution (Lu et al., 2011). Its performance against corrosion was evaluated in 3 wt% NaCl solution by electrochemical impedance spectroscopy (EIS) and anodic polarization. The catalytic effect of PAN was responsible for the formation of dense oxide layer at the metal/PAN interface. The role of dopant in enhancing the performance of PAN coatings was indicated in many cases. For instance, PAN-molybdate coating prepared on a steel substrate via cyclic voltammetry exhibits improved protection performance in 1% NaCl solution. On the basis of EIS and potentiodynamic polarization results, the formation of Fe–molybdate complexes, besides passivation, was considered responsible for the enhanced protection (Karpakam et al., 2011). In the case of PAN-benzoate coatings, galvanostatically synthesized on mild steel, it was suggested that the protection effect of PAN-benzoate on mild steel using a partial PAN coating can be described by the "switching zone mechanism" (Elkais et al., 2013). According to this mechanism, during the initial stage of the exposure of the partially protected mild steel to 3 wt% NaCl, the corrosion potential shifts to more negative values determined cathodically by the dedoping process of PAN-benzoate along with the oxygen reduction and anodically by the iron dissolution. As most of the benzoates were released, the PAN conductivity decreases. During intermediate times of exposure, the corrosion potential shifts to less negative values determined by the cathodic oxygen reduction at the bare metal in the bottom of the PAN pores and the anodic partial doping of PAN by by chlorides. Then the initial stage restarts, and the cycle is repeated. It is suggested that during a prolonged exposure of the partially coated mild steel, a thin oxide layer is formed near the PAN, and thus, the oxygen reduction is catalyzed.

Polypyrrole: Electrodeposition of PPy in a one-stage process on Fe, Al, Cu, and SS has been carried out from aqueous and organic media (Beck & Michaelis, 1992; Ferreira et al., 1996; Gonzalez & Saidman, 2011; Herrasti et al., 2007; Hulser & Beck, 1990). PPy can be deposited in different ways, such as a simple film (Nguyen Thi Le et al., 2001), as part of a multilayered coating, or incorporation as an additive in a matrix (Lacaze et al., 2010). A PPy coating can act as a remarkable corrosion inhibitor for metal substrates, although the adhesion of the polymer to the metallic substrate is frequently poor if either a specific surface pretreatment was not applied or suitable dopants are not used during the synthetic procedure. Protection time was found to increase in an exponential way on the charge deposited galvanostatically. The impedance spectra can be rather ascribed to PPy films than to the underlying passive film, and the protection mechanism was mostly of galvanic type.

The presence of different ions in the electrolyte solution during the electrochemical synthesis of CPs produces coatings with certain characteristics and properties. An improvement of the corrosion protection of 55% Al–Zn-coated steels by PPy was found when molybdate is inserted into a PPy layer deposited galvanostatically in tartrate solutions of different pHs (Ryu et al., 2012). A recent investigation demonstrated that PPy films electrosynthesized from an alkaline solution of pH 12 containing nitrate and molybdate onto AISI 316L SS have good adhesion and can completely protect the substrate against pitting corrosion in chloride solutions, even when the sample has been polarized at very positive potentials during a considerable period of time (Gonzalez & Saidman, 2011).

PPy doped with corrosion-inhibiting ions has been studied for the corrosion protection of metals and their alloys. Molybdate- and nitrate-doped PPy was electropolymerized on the surface of AISI 316L SS in neutral and alkaline media (Gonzalez & Saidman, 2011). It was found that the pitting potential of the substrate was reduced. Improved corrosion protection was attributed to the electrochemical activity of PPy, the formation of a passive oxide layer, and the nature of dopants used in the electrochemical synthesis of PPy on the substrate. In another study, electrochemical polymerization of PPy was carried out on AA 2024-T3 substrate using various dopants, such as camphor sulfonic acid, phenylphosphonic acid, para-toluene sulfonic acid, oxalic acid, and cerium nitrate salt. Different morphologies and varying corrosion behaviors of the PPy were observed as a function of dopant anion (Balaskas et al., 2011).

It was determined that the $FeC_2O_4.2H_2O$ coating was first deposited on the steel substrate, resulting in its passivation. However, the $FeC_2O_4 \cdot 2H_2O$ layer was decomposed when the electropolymerization potential of Py was reached. Many of the important properties of the resultant coating, such as adhesion, were found to be dependent on local pH and applied current densities. It was further demonstrated in this study that the electrodeposited PPy coating provided better corrosion resistance as compared with the electrodeposited poly(N-methylpyrrole) coating (Su & Iroh, 2000).

Several researchers have attempted to electropolymerize Py on different substrates other than steel and aluminum. Herrasti et al. (2007) electrodeposited Py on Cu. The deposited layer had low porosity and homogeneous surface characteristics. This was the outcome of the constant growth rate of the PPy films on Cu surfaces. These PPy deposits, which were formed on Cu, showed excellent corrosion resistance in 3 wt% NaCl solution. They also showed that a higher monomer concentration of about 0.3 M should be used to have good barrier layer characteristics and good redox properties (Herrasti et al., 2007). Jiang et al. (2003) electrodeposited yellow-black PPy coatings on AZ91 magnesium alloys in alkaline solutions. They found that the different pretreatments have an effect on the characteristics of potentiodynamic curves, thereby changing the resultant growth rate. Tuken et al. (2004) attempted electropolymerization of Py on the surface of brass and Cu. As the resulting PPy films have better adhesion on the oxidized Cu, better corrosion protection was offered as compared with PPy-free brass. They proposed a protection mechanism based on barrier effects exhibited by these coatings. Protection of Cu by PPy was also investigated by using phytic acid ($C_6H_{18}O_{24}P_6$) solutions for the electrodeposition of PPy on Cu surfaces (Lei et al., 2013). Phytic acid (IP_6), as the principal phosphorous storage form in many plant issues, is a green material that forms a complex with Cu ($Cu_n(C_6H_{12}O_{24}P_6))^{(6-2n)-}$. The formation of the complex was supported well by surface analysis using X-ray photoelectron spectroscopy (XPS) along with infrared reflection absorption spectroscopy (IR-RAS) spectroscopy. It was shown that the dissolution of Cu is greatly inhibited for a long period in 3.5 wt% NaCl solution. However, PPy-IP_6 was degraded with a long immersion because of the gradual loss of the oxidative property of PPy, as was indicated by the increase in polarization resistance and the shift of the E_{OC} to more negative values (Figure 4.8). Doping of Na^+ dominates instead of doping of IP_6.

$$PPy^{(x-y)+} - (x/6)IP_6^{6-} - yNa^+ + ye^- - \prod PPy^{(x-y)+} - (x/6)IP_6^{6-} yNa^+$$

PPy is a very attractive organic coating for a large number of biological and biomedical applications owing to its biocompatibility with the human body (Wallace et al., 2003). It was demonstrated that a PPy film was capable of protecting NiTi alloy against localized corrosion when it was synthesized in a neutral solution of sodium bis(2-ethylhexyl) sulfosuccinate containing Py (Flamini, 2010). Several corrosion inhibitor ions such as molybdate and citrate have been used to passivate the surface of Ti and Ti_6Al_4V alloy. Furthermore, molybdate ion was used as a corrosion inhibitor of NiTi alloy over a wide range of chloride ion concentrations. In this case, potentiodynamic curves of the NiTi alloy passivated in molybdate solution show a positive displacement of 0.4 V_{SCE} in the breakdown potential value when compared with the nonpassivated sample. This result was explained by an increase in TiO_2 content on

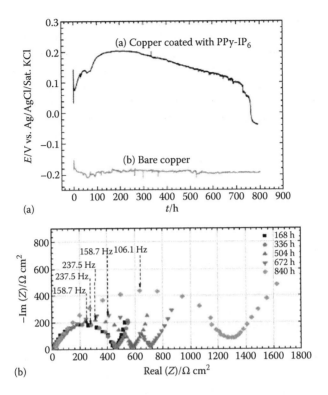

FIGURE 4.8 **(See color insert.)** (a) Monitoring of the open circuit potential of the Cu–PPy-IP$_6$ electrode in comparison with the bare Cu electrode with time in 3.5 wt% NaCl. (b) A decrease in the oxidative ability of the PPy-IP$_6$ layer results in characteristic changes in the Nyquist spectra, with time indicating an increase in the polarization resistance and an increase in the capacitance. (From Lei, Y.H., Sheng, N., Hyono, A., Ueda, M., Ohtsuka, T., *Corros. Sci.*, *76*, 302–309, 2013.)

the outermost surface (Flamini et al., 2014). The TiO$_2$-containing PPy-molybdate coating has features of an advanced protective coating that can completely protect the substrate against pitting corrosion in 0.15 M NaCl solution, even when the coated sample is polarized during a prolonged period to a potential higher than the breakdown potential of the bare substrate (0.65 V$_{SCE}$). It seems that more than one protection mechanism works simultaneously to offer this enhanced protection to NiTi alloy. The possible causes of the corrosion protection properties of the coating are as follows: (1) the presence of the molybdate anion as corrosion inhibitor modifies the properties of the passive oxide layer, giving the substrate greater adherence to the polymer because of the formation of a composite structure of TiO$_2$/PPy; (2) the PPy film as an electroactive polymer induces a galvanic interaction with the substrate; (3) the fixed negative charge of molybdate ions prevents the ingress of chlorides into the polymer matrix; and (4) the release of corrosion inhibitor anions from the polymer matrix also contributes to the protection of the alloy.

As was shown by the data reported in the literature and the examples mentioned previously, the protective properties of PPy primers cannot be maintained for very long periods in chloride-containing solutions because of the degradation of PPy since it loses its oxidative ability. Degradation of PPy is also observed in alkaline media. The impact of environmental factors (i.e., pH, temperature, dissolved oxygen, electrolyte concentration), electrochemical stimulation, and doping anions on the protective behavior of PPy films was examined in alkaline aqueous media by cyclic voltammetry (Qi et al., 2012). When a PPy-A film (A stands for the dopant anion) is placed in alkaline aqueous solutions, OH− may exchange A− ions, and/or may attack the PPy-A film due to the strong nucleophilic character of OH−. Therefore, PPy/A films may corrode by two ways, as the reaction schemes of Figure 4.9 shows. The higher the concentration of OH−, the more severe the PPy/A degradation.

Polythiophene: PTh-based coatings were employed to a lesser degree for prevention. Despite the limitations mentioned previously for the electrochemical polymerization of Th, electrodeposition of PTh leading to the formation of protective coatings on oxidizable metals (mild steel) was carried out from acetonitrile solutions (Aeiyach et al., 1997; Kousik et al., 2001). EIS studies support enhancement of the passivity of the PTh-coated mild steel. Water uptake and delaminating area studies also confirmed the protective action of electropolymerized PTh on the mild steel surface (Kousik et al., 2001).

The problem with the relatively high oxidation potential of Th was addressed in early studies by using bis-thiophene (E^0 ~ 1.8 V$_{SCE}$) in aqueous-organic media during electrodeposition of PTh on oxidizable metals (Barsch & Beck, 1993). In fact, the first report on the electrodeposition of PTh on iron from bisthiophen-containing aqueous-organic media of oxalic acid or KNO$_3$ led to thin layers (~0.1–3 μm) that do not dissolve and are stable in the air. Certainly, overcoating with paint is required, although complications may arise because of the electrochemical interaction with the substrate through the pores that gradually will result in degradation of the PTh interlayer. Moreover, it was shown that electropolymerization of bisthiophene on iron can be carried out galvanostatically from aqueous micellar systems containing SDS as a surfactant, a small amount of butanol, and LiClO$_4$ as a supporting electrolyte (Bazzaouri et al., 1996).

As was mentioned previously, 3-substituted thiophenes were found more appropriate than Th in preparing protective coatings against metal corrosion. Poly(3-methylthiophene) (P(3-MT)) electrodeposited from acetonitrile solutions on iron surfaces where adsorption of 2-(3-thienyl) ethylphosphonic acid first occurred shows improved adhesion on the substrate as compared with the Fe/P(3-MT) electrodes prepared without this surface pretreatment. Delamination tests were carried out in the presence of an alkyd layer acting as top coat (Rammelt et al., 2001).

More recent research on PTh include poly(3,4-ethylenedioxythiophene) (PEDOT) based composite materials. PEDOT is among the most promising PTh derivatives to be used in corrosion protection at various coating formulations owing to its higher conductivity and significant stability as compared with PTh. PEDOT-based composite coatings were electrodeposited using CPS deposition on GC and SS (X20Cr13) substrates in the presence of 4-(pyrrole-1-yl) benzoic acid (PyBA) and phosphododecamolybdic acid (PMo12) (Adamczyk & Kulesza, 2011). Compact and tight composite coatings of PEDOT/PyBA/PMo12 have been successfully generated electrochemically on the metal substrates. The coating growth was facilitated in the

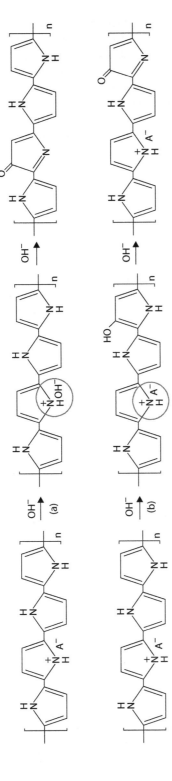

FIGURE 4.9 Two ways by which the strong nucleophilic OH⁻ groups may attack the PPy⁺-A⁻ films. First, the OH⁻ ions are exchanged by dopants A⁻, and then PPy⁺-OH⁻ is overoxidized to a Py ring with carbonyl groups, resulting in nonstability, and second, the OH⁻ ions attack the Py ring directly to form carbonyl groups. In this case, the PPy⁺-A⁻ film will lose its conjugated double-bond structure and electrochemical activity. (From Qi, K., Qiu, Y., Chen, Z., Guo, X., *Corros. Sci., 60,* 50–58, 2012.)

presence of polyoxyethylene-10-laurylether (BRIJ10) neutral surfactant. The coatings have exhibited very good adhesion to the substrate and ensured fairly effective protection against pitting corrosion in a strongly acidic chloride-containing medium.

The PEDOT/PyBA composite coating serves as a stable host matrix for the large phosphomolybdate anions. The resulting coating was denser in structure, and due to the existence of electrostatic repulsion effects, the access of pitting-causing anions (i.e., chlorides) to the surface of SS was largely prevented. The excellent adhesion of the coating onto the SS substrate offers interfacial stabilization to the SS/PEDOT/PyBA/PMo12 system. Indeed, as Figure 4.10 shows, the cyclic voltammetric response of the SS/PEDOT/PyBA/PMo12 electrode is very close to the corresponding response obtained by the GC/PEDOT/PyBA/PMo12 electrode, indicating the excellent stability of the PEDOT hybrid coating. This originated perhaps from the possible interactions of phosphomolybdates with metal ions, such as Cr(III), Fe(III), or Fe(II), that exist at the SS surface. The PEDOT/PyBA/PMo12 coating considerably extends the induction time required for pit nucleation on SS, when compared with the PyBA-free PEDOT/PMo12 coating.

Furthermore, incorporation of heteropolyacids, $H_3SiMo_{12}O_{40}$ ($SiMoO_{12}$) along with the PyBA was demonstrated during the electropolymerization of PEDOT in aqueous media (Adamczyk et al., 2014), which led to another stable SS/PEDOT/PyBA/SiMo12 electrode with excellent adhesion of the SiMo12-containing PEDOT coating to the SS substrate. The very good adhesion to the substrate ensures a fairly effective protection against pitting corrosion in a strongly acid medium containing chloride anions, as can be seen in Figure 4.11. The corrosion potential, E_{corr}, shifts to more noble potentials, whereas the electrodissolution currents decrease by several orders of magnitude.

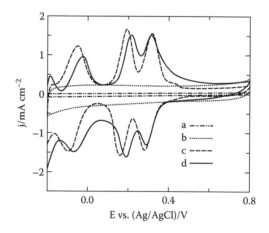

FIGURE 4.10 Voltammograms of the PEDOT/PyBA/PMo12 coating on glassy carbon (d). For comparison, the responses of PEDOT (a), PEDOT/PyBA (b), and PEDOT/PMo12 (c) coatings are provided. Electrolyte: 0.5 M H_2SO_4. Scan rate: 50 mV s^{-1}. (From Adamczyk, L., Kulesza, P.J., *Electrochim. Acta*, 56, 3649–3655, 2011.)

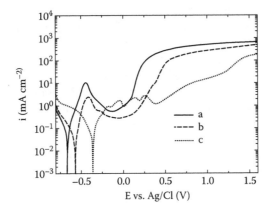

FIGURE 4.11 Potentiodynamic polarization curves for bare SS (a) and for SS specimens coated with PEDOT/PyBA (b) and PEDOT/PyBA/SiMo12 (c). (From Adamczyk, L., Giza, K., Dudek, A., *Mat. Chem. Phys.*, *144*, 418–424, 2014.)

Polycarbazole: Studies on the anticorrosion properties of carbazole derivatives are rare (Dudukcu et al., 2009; Duran et al., 2013; Frau et al., 2010). The electrodeposition and electrochemical grafting of a carbazole containing precursor polymer (copolymer) with an "active" conjugated polyfluorene backbone (Figure 4.12) was carried out onto of AISI 304 SS type from acetonitrile solutions, resulting in conjugated polymer network (CPN) (Frau et al., 2010). Polycarbazole (PCz) films, electrochemically synthesized on SS by this precursor approach, act as protective coatings, and they reduce the corrosion rate of steel coupons by providing anodic protection. It was found that the corrosion performance of these coatings depends on the type of the supporting electrolyte, and hence the dopant ion, used in electrosynthesis.

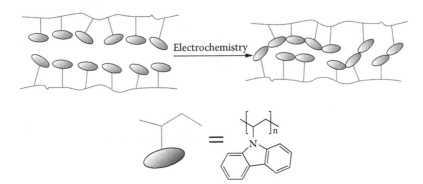

FIGURE 4.12 Schematic representation of the precursor polymer approach to CPN films, according to which, the precursor polymer has an electroactive unit [(poly(vinylcarbazole)], which can be electrochemically polymerized to give an extended π-CPN through both intermolecular and intramolecular connectivity. (From Frau, A.F., Pemites, R.B., Advincula, R.C., *Ind. Eng. Chem. Res.*, *49*, 9789–9797, 2010.)

The protective properties of PCz films against corrosion of AISI 304 SS seem to be correlated with the size of counter cation of the dopant (Duran et al., 2013). Larger cations result in a lower conductivity of the resulting film but enhance the barrier property of the film against ingress of corrosive species. On the other hand, corrosion tests indicated that, besides conductivity, film thickness and morphology are also effective on the protective performance of PCz films. An example of the redox behavior of the PCz traced during the electrochemical polymerization of 0.01 M Cz from acetonitrile solutions containing 0.1 M tetrabutylammonium perchlorate (TBAP) is depicted in Figure 4.13a. The morphology of the PCz-coated SS samples was examined by scanning electron microscopy (SEM). The SEM micrograph of the SS/PCz(TBAP) surface can be seen in Figure 4.13b. Among three supporting electrolytes used, TBAP, sodium perchlorate (SP), and lithium perchlorate (LP), it was found that TBAP provided the best protective efficiency in the order PCz(TBAP) > PCz(SP) > PCz(LP). These PCz thin films are of interest for the protection of steel bipolar plates in the proton exchange membranes fuel cells.

The presence of Cz ring was found also to play a crucial role on the performance of poly(N-vinylcarbazole) (PNVCz), poly(N-vinylcarbazole methylethylketone-formaldehyde resin) (P[NVCz-MEKF-R]), carbazole methylethylketone formaldehyde resin (Cz-MEKFR), and poly(carbazole methyl ethylketone formaldehyde resin) (P[Cz-MEKFR]) coatings when used to improve the photoactivity of pyrite, which is needed for solar cell applications by protecting the pyrite from corrosion and photocorrosion. Protection can be achieved by passivation of the pyrite before formation of the polymer layer and also acting as barrier against corrosion in contact with solutions after formation of the polymer film (Ustamehmetoglu et al., 2013).

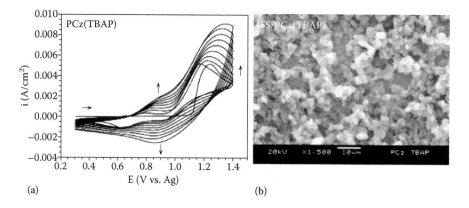

(a)   (b)

FIGURE 4.13   (a) Cyclic voltammograms obtained at dE/dt = 50 mV s^{-1} during the electrodeposition of PCz on SS from acetonitrile solutions containing 0.01 M Cz and 0.1 M TBAP. (b) SEM micrograph of PCz-coated surfaces. (From Duran, B., Çakmakcı, İ., Bereket, G., *Corros. Sci.*, 77, 194–201, 2013.)

4.3.1.2   Casting-Based CP Coatings

Casting of CPs is a multiple step process and includes (1) dissolving or dispersing of a CP in a suitable solvent, (2) spreading it over a metal surface, and (3) evaporation of the solvent to leave behind a film. A major problem is that a suitable solvent is required, in particular in the case of PAN. Several approaches have been suggested to overcome insolubility and unprocessability problems, including chemical modifications of the CP composition by functionalization via insertion of substituents, suitable to render solubility in given medium and/or by covalently bonded dopants (Wei & Epstein, 1995; Yue et al., 1992), mostly with sulfonic acids acting as internal dopants. Attachment of sulfonate in the polymer backbone can be achieved by different methods (Koul et al., 2001; Wei et al., 1996; Yue et al., 1991, 1992) instead of attaching it simply as a dopant (Camalet et al., 1998b; Ozyilmaz, 2006). Besides the self-doping property of sulfonated PAN, its enhanced crystallinity is also considered as a beneficial factor for its protective performance in HCl against pitting corrosion of iron (Koul et al., 2001).

Polyaniline: The formation of casting films is mostly used for chemically synthesized CPs. A common solvent for PAN dissolution is *N*-methyl-pyrrolidone (NMP), which in fact leads to gelation. PAN transmitted from NMP in the emeraldine base (EB) state and stabilized on Fe surfaces exhibits better protective properties than does poly(vinyl chloride) (PVC) in 3.5% w/w NaCl solution (Mirmohsen & Oladegaragoze, 2000). It was observed that PAN films formed in NMP in the undoped form, when in contact with chloride acidic solutions, undergo a nonuniform doping process, resulting in an increase in its conductivity (Santos et al., 1998). In particular, chemically prepared PAN in its undoped EB state was dissolved in a solution 2 wt% of NMP, sprayed on plates of carbon steel (CS) and AISI 304 SS, and then dried in an oven at 45°C for 12 h. The PAN films prepared by this procedure presented a strong adherence to the metallic substrate. Potentiodynamic measurements showed that the protective action of casting PAN promotes a positive shift of the corrosion potential for CS (~100 mV) and for SS (~270 mV) owing to the formation of a passivated layer that blocks the corrosion progress. The inhibition efficiency for the corrosion processes, evaluated by weight loss measurements, was almost 100% for both kinds of steel. It is noteworthy that while the PAN film loses water when kept out of the solution, it returns to the original state after some time in contact with the solution without losing its electrical and mechanical characteristics.

Besides NMP used to prepare dispersions of PAN in the undoped EB state, which seems to undergo doping when in contact with an aqueous acid medium with or without chlorides, xylene or chloroform was used as solvents for different forms of PAN. A comparison was carried out between two techniques of casting on mild steel by using (1) undoped PAN from NMP and then doped with acid and (2) doped PAN directly obtained from xylene or chloroform (Pud et al., 1999). A higher decrease in corrosion current was observed in case 1 as compared with case 2.

Insertion of sulfonic acid groups to the PAN molecular structure carried out by chemical copolymerization of AN with *m*-aminobenzenesulfonic acid (metanilic acid, MA) at various ratios of AN:MA led to partially self-doped poly(aniline-co-metanilic acid) (PANMA) with different sulfonation degrees. Dedoped forms of PANMA were dispersed in ethanol (5.0 mg/L), and the dispersion was drop-cast onto a polished CS surface and dried at room temperature to form a thin coating (Xing et al., 2014). Potentiodynamic polarization and EIS measurements showed that PANMA coatings provide good anticorrosion protection on CS in I M H_2SO_4. It was found that low sulfur content in the PANMA coatings (lower degree of self-doping by the -SO_3H group) afforded the best protection, although the performance of PANMA coatings was inferior to that of PAN itself under the applied conditions. It is suggested that this behavior might be attributed to the hydrophilic properties of the sulfonic acid group. -SO_3H. The degree of hydrophilicity or hydrophophobicity of metallic surfaces is a critical factor for their susceptibility to corrosion. However, self-doping is anticipated to be advantageous in chloride-containing corrosive solutions preventing anion exchange with the small chloride ions. This is an open question, and more studies are required toward the optimization of the properties of CP-based coatings in relation with the existing corrosive conditions.

Polypyrrole: Direct electrochemical deposition of PPy is the most utilized method for PPy-based coatings. The relatively low oxidation potential and the noncarcinogenic characteristic of the monomer Py, as compared with AN, are favorable for selecting the electrodeposition method. In addition, Py polymerization in neutral and alkaline solutions where electrodissolution rates of active metals essentially decrease is feasible. Electrochemical synthesis of PPy is mostly used also in the case of indirect deposition of PPy-based coatings (Wallace et al., 2003).

Polythiophene: Several derivatives of Th, such as P(3-MT), were examined with respect to their anticorrosive properties utilizing the casting method. Poly(3-octyl thiophene) (P3-OT) and poly(3-hexylthiophene) (P3-HT) dissolved in toluene were deposited by drop casting onto 1018-type CS and their protective ability against the corrosion of CS was tested in 0.5 M H_2SO_4 (Medrano-Vaca et al., 2008). Two surface finishings of CS were examined, namely, CS abraded with 600-emery paper and with alumina (Al_2O_3) particles of 1.0 µm in diameter (mirror finish). Their corrosion resistance was estimated by using electrochemical measurements. In all cases, polymeric films protected the substrate against corrosion, but the protection was improved if the surface was polished with Al_2O_3 and was better in the case of P3-HT because the amount of defects was much lower as compared with P3-OT films. It is suggested that both P3-HT and P3-OT films did not act only as a barrier layers against aggressive environment, but they improved the passive film properties by decreasing the critical current required to passivate the CS and by broadening the passive potential region.

4.3.2 Paint Blended Coatings

The use of CP blends is widely explored with polymers such as epoxy and acrylic resins or polyurethane (Baldissera & Ferreira, 2012; Deshpande et al., 2014; Khan et al., 2010; Sathiyanarayanan et al., 2006a,b, 2008; Souza, 2007; Torresi et al., 2005). The properties of blends depend significantly on the CP, polymer/solvent matrix, dopant, and preparation conditions. Several methods of preparation have been developed. CP mutliblends can be made by codissolving, for instance, PAN and an appropriate matrix polymer (i.e., poly(methyl methacrylate)) in a common solvent (i.e., *m*-cresol) and processing the CP blend directly from liquid mixture. The multiblend mixture can be processed to produce coatings on substrates of excellent electrical and optical features.

PAN-based paints result in coatings of high corrosion resistance for steel surfaces. (Baldissera et al., 2012; Baldissera et al., 2010; Iribarren et al., 2005; Sathiyanarayanan et al., 2006b). As far as the redox catalytic activity of PAN can function to tolerate small sized defects and the steel passivity is retained (Wessling, 1994). In the presence of large defects PAN coating may fail to provide protection. It was shown that the dopant anions play a decisive role in this deterioration mechanism by which PAN and other type of CP-based coatings lose its protection ability (Rohwerder, 2009). To decrease transport of dopant anions sulfonated PAN (Baldissera & Ferreira, 2012), and large sized dopants such as dodecylbenzenesulfonic acid [DBSA] (Baldissera et al., 2010), camphorsulfonic acid (CSA) (Souza, 2007; Silva et al., 2005, 2007) and p-toluenesulfonic acid (PTSA) (Kinlen et al., 1999, 2002) in PAN blends. PAN-polyvinyl blends doped with different sulfonic acids can be prepared by in situ dispersion polymerization (Bhadra et al., 2014). The resulting blends exhibit good thermal stability and conductivity and this might explain their suitability for corrosion protection.

The possible scheme of the hydrogen bonding between PAN-polyvynil alcohol (PVA) and acids CSA, p-napthalenesulfonic acid (NSA), DBSA and PTSA and the steps for doped PAN-PVA blends are shown in Figure 4.14 (Bhadra et al., 2014). The thermal stability of the DBSA-doped blend was found to be better than that of the other three acids. Differential scanning calorimetry results establish that the obtained polymer blends are miscible. The DBSA-doped PAN blend exhibits the maximum DC conductivity. The conductivity is found to be highest for the DBSA-doped blend, and this result could be explained by the long tail alkyl surfactant nature of this acid.

Polyaniline: PAN doped with benzenesulfonate (BS) and lignosulfonate (LS) was also synthesized by chemical oxidative polymerization and was incorporated into a chlorinated rubber binder in varying weight proportions (Sakhri et al., 2011). The coatings were applied on mild steel and the corrosion performance behavior was monitored by EIS measurements with salt spray exposure and immersion in NaCl solution. Below 1% concentration of BS-doped PAN and 3% concentration of LS-doped PAN showed improved corrosion performance. It was observed that LS was not covalently bonded to PAN, which was responsible for the different behavior as compared with that of the BS-doped PAN. PAN-LS/epoxy coatings were also employed for corrosion protection of AA2024-T3 (Gupta et al., 2013). The performance of PAN-LS/epoxy blends was investigated in 0.6 M NaCl during 30 days for different loadings in PAN-LS. Better protection efficiency was found for an optimum amount of PAN-LS (~5 wt%) in epoxy blends. It seems that at these optimized concentrations of PAN-LS, a uniform distribution of PAN particles results in a decrease in the permeability of H_2O and O_2. The improved anticorrosion performance is explained in terms of the anodic protection mechanism and the controlled inhibitor release mechanism. A thickened oxide film is formed because of the active redox role of PAN, whereas the release of dopant (LS) at the defect site prevents

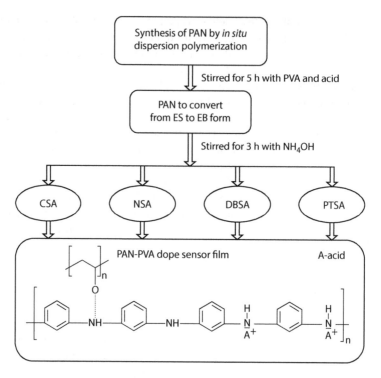

FIGURE 4.14   Flow diagram for the synthesis of PAN-PVA thin films. Inset: The scheme of hydrogen bonding between PAN-PVA and acids, CSA, NSA, DBSA, and PTSA. (From Bhadra, J., Madi, N.K., Al-Thani, N.J. Al-Maadeed, M.A., *Synth. Met.*, *191*, 126–134, 2014.)

active dissolution of the substrate by the formation of an aluminum–sulfonate complex. Moreover, DBSA-doped PAN nanoparticles (1 wt%) blended epoxy ester (EPE) showed better corrosion protection of the CS substrate in 3.5 wt% NaCl solution compared with the simple EPE coating (Arefinia et al., 2012). This behavior was also attributed to the released dopant anions that, with the iron cations, provide a secondary barrier layer, which passivates the CS.

Polypyrrole: The corrosion protection of PPy and PAN coatings, electrochemically deposited with and without PVA as adhesive onto AISI 304 type SS, has been evaluated in 0.5 M H$_2$SO$_4$ at 60°C using electrochemical techniques (potentiodynamic polarization curves, potentiodynamic polarization and electrochemical impedance spectroscopy (EIS) measurements. Results showed that the open circuit potential of the substrate was shifted to positive values up to 500 mV with the polymeric coatings. The corrosion rate was lowered by using the polymers, but with the addition of PVA, it was decreased further, one order of magnitude for PPy and up to three orders of magnitude for PAN. Impedance spectra showed that the corrosion mechanism is controlled by a Warburg-type diffusional process of the electrolyte throughout the coating and that the uptake of the environment causes the eventual failure of the coating corroding the substrate (Tan & Blackwood, 2003).

Polythiophene: It has been observed that the derivatives of PTh are less complicated than the derivatives of other CPs such as PAN and PPy. It is this feature of PTh that makes easy for researchers to synthesize a variety of substituted polymers. Some of the derivatives of PTh present high conductivity values and excellent performance as compared with other CP derivatives. Therefore, the principal method used for the formation of PTh-based coatings is by using electrodeposition techniques. By applying suitable potential/current values, PTh and its derivatives can be generated as thin adherent films on metal surfaces or on other CP substrates like PPy. For example, deposition of PTh was carried out on a PPy-coated Cu electrode from acetonitrile containing 0.1 M Th and LiClO$_4$. The combination of PPy/PTh has resulted in significant corrosion protection of Cu surfaces for relatively prolonged immersion (Tuken et al., 2005). An alternative procedure to form adherent PTh films is by using an adhesion promoter (AP) consisting of an anchor group (phosphono acid) such as 2(3-thienyl) ethyl phosphono acid. Rammelt et al. (2001) investigated the role of such special surface treatment on mild steel. A schematic presentation of the film formation of P(3-MT) films on pretreated mild steel is seen in Figure 4.15.

The AP is covalently bound to the steel substrate and an alkyl spacer with a Th unit is at the end of the spacer. The AP layer provides an easy way for functionalizing the steel surface and for surface structuring. The layer thickness is reinforced with an oxidation process in the presence of methylthiophene units. The AP layer can be considered as a self-assembled monolayer (SAM), which transforms the SAM structure to the first layers of the polymer film. The P(3-MT) film is more ordered than P(3-MT) deposited directly on the steel surface without pretreatment. Homogeneous and very adherent P(3-MT) films were formed on the surface of mild steel. These films of P(3-MT) were ultrathin (thickness was ~1 µm), highly ordered and exhibited better protection, contributing to separating effectively the electrochemical processes of oxygen reduction and iron dissolution in the surface region (Rammelt et al., 2001). This rational of AP incorporation can be very easily used for a well-ordered build-up of different homogeneous CP films with improved corrosion protection.

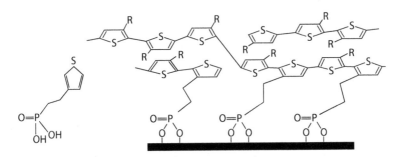

FIGURE 4.15 Schematic representation of the role of an AP on the formation of adherent films of P(3-MT) films. (From Rammelt, U., Nguyen, P.T., Plieth, W., *Electrochim. Acta, 46*, 4251–4257, 2001.)

4.3.2.1 CP Additives in Marine Paint Formulations

CPs are used as anticorrosive additives to modify the formulation of conventional organic coatings (paints) (Iribarren et al., 2004). A very low concentration (0.2–0.3 wt%) of CP is required to offer protection to the metal substrate. PAN (Armelin et al., 2008; Gzgur et al., 2015; Iribarren et al., 2005; Sathiyanarayanan et al., 2010), PPy (Armelin et al., 2008), PTh derivatives (Ocampo et al., 2005) have been utilized to modify common paints such as epoxy, alkyd, and polyurethane coatings. Homogeneous emulsions of the modified paint are usually prepared by a previous dispersion of the CP in suitable solvents followed by mechanical stirring with organic coating. It was shown that the characteristics of the modified paints are frequently very similar to those of the original paint. Although, in most cases, the CP-modified paint keeps its mechanical and thermal stability, viscosity, and other physicochemical properties, several exceptions were observed. For example, the epoxy coating modified by a low concentration of PPy composite with carbon black presents a brittle failure with higher modulus of elasticity than that of the paint alone (Armelin, et al., 2009), while poly(3-decylthiophene-2,5-diyl) used at 0.2%–0.3% w/w changes the color of the paints to violet (Iribarren et al., 2004; Ocampo et al., 2005). Moreover, an epoxy-based coating modified by the addition of PEDOT electrochemically synthesized exhibited important improvement in the protection (Armelin et al., 2007b; Ocampo et al., 2005). The corrosion resistance of the PEDOT-modified epoxy coating was increased by increasing the doping level in support of the active role of CP additive in organic paint, regardless of its relatively very low concentration.

The advantage of these CP formulations is that very low concentrations of CP offer efficient active corrosion in combination with the advanced properties of the paint regarding thermal and mechanical stability, adhesion to the substrate, homogeneity, and barrier properties. The protective properties imparted by the CP-additive depend on the following:

1. The paint formulation. For instance, polyurethane was not favorable for PPy pigment, in contrast to acrylic and polyester resin (Armelin et al., 2007b).
2. The PAN state. The protective performance of PAN in its emeraldine salt (ES) state as an additive of alkyd resin was improved against the corrosion of the steel in comparison with that of the control alkyd coating, while the PAN-ES-modified alkyd resin presented increased degradation resistance (Iribarren et al., 2005). In contrast to the PAN-ES-free coating, the PAN-ES in the modified alkyd coating prevents the corrosive species from reaching the metal surface and, hence, a direct attack of oxygen and water to the steel substrate. Corrosion tests indicated that this polymer might work as both a corrosion inhibitor and an AP. Improved anticorrosion performance was also observed when using coatings constituted by either PAN-ES or PPy (composite with carbon black) as additives of an epoxy paint coating (Armelin et al., 2008). Immersion tests in an aggressive solution showed better corrosion resistance for the paints modified by adding PAN and PPy than for the control epoxy coating.

It is worth noting that by adding PAN in its EB state, the protective properties of the epoxy paint are not altered. Three paints, used as primers in marine environments, were checked: two epoxy coatings that differ in the presence or absence of inorganic anticorrosive pigment (zinc) and one alkyd coating. Experiments showed that the PAN-EB did not affect the protective properties of the epoxy without inorganic anticorrosive pigment nor the alkyd formulations. On the contrary, the PAN-EB added to the epoxy paint with inorganic anticorrosive pigment induced the formation of a zinc oxide layer, which promoted the corrosion attack (Armelin et al., 2007a).

3. The concentration of the CP additive. In the case of the PPy composite employed as anticorrosive additive with a concentration 1 wt%, the corrosion resistance of the epoxy paint improves with respect to the control sample during 480 h of exposure in saline solution. The addition of a larger concentration of PPy to the paint, i.e., 0.3 wt%–1.5 wt% produced only a small improvement with respect to the unmodified coating.

4. The dopant. The protection properties depend, besides the oxidized or reduced state of the PAN additive, on the dopant ion. Phosphate- and chloride-doped PANs were synthesized by chemical polymerization methods and used as additives in epoxy–coal tar coatings. The protective capability of the coatings has been studied by open circuit potential (OCP) and EIS measurements in 3 wt% NaCl solution. Coatings containing 1 wt% and 3 wt% phosphate-doped PAN and 3 wt% chloride-doped PAN were found to be highly corrosion resistant. The higher corrosion protection ability of phosphate-doped PAN is a result of the redox property of PAN along with the formation of iron–PAN complex and secondary layer of iron-phosphate layer on steel (Sathiyanarayanan et al., 2009). By a similar mechanism, the enhanced protection property of sulfonate-doped PAN pigment was also explained. The sulfonate-doped PAN containing paint was prepared using vinyl resin (molecular weight [Mwt.] 30.000) with 1 wt% PAN as pigment. The paint was applied on the sand blasted (Sa 2.5) mild steel panel and evaluated after 10 days of curing at room temperature. The coating thickness was 50 ± 5 μm (Sathiyanarayanan et al., 2010). It is found that the sulphonate-doped PAN containing vinyl coating offers more corrosion resistance in both acid and neutral media than the vinyl coating does. FTIR studies have shown that the conducting state of PAN in the coating is not changed in acid media, while PAN is changed to its nonconducting form in neutral media. The mechanism of corrosion protection is found to be the formation of PAN–iron–sulphonate complex beneath the coating, along with the formation of the passive iron oxide film.

Zaarei et al. (2012) suggested a new way of dispersing PAN-EB in the hardener amine and curing the epoxy resin-PAN-EB mixture to avoid phase separation between the PAN-EB and epoxy resin matrix The aminic hardener was prepared by dispersion of EB-PAN in 3 (aminomethyl)-3,5,5-trimethylcyclohexan-1-amine employing sonication, centrifuging, and submicron filtering methods. The presence of 0.5 wt% EB in initial mixture of EB-hardener compositions was sufficient to lead to coatings with relatively better anticorrosion protection for steel compared with neat resin coating. The presence of initial 2.5 wt% of EB in the hardener resulted in the formulation of an epoxy coating with superior corrosion protection properties.

The properties of the cured epoxy resins depend on their structure, the extent of cure, the reaction kinetics such as the curing conditions, the time, and the temperature of cure. To obtain high-performance resin, the relationships between the structure of the networks and the final properties should be understood (Fu et al., 2009). Based on FTIR spectra, the mechanism for epoxy resin curing PAN was postulated as self-curing reactions (Figure 4.16) by the reactions between amine groups of PAN with epoxide groups of epoxy resin (Siva et al., 2014).

By entrapping a small amount of PAN-EB such as 0.5 wt% in the epoxy resin matrix, improved corrosion protection of steel in 3 wt% NaCl solution was observed compared with the neat epoxy resin coating (Siva et al., 2014). It is found that the coating has an initial resistance of $4.7 \times 10^7 \, \Omega \, \text{cm}^2$, which is nearly two orders higher than that of conventional amide cured epoxy coating, and with time, it decreases slowly. During the period of corrosion evaluation in 3 wt% NaCl, there

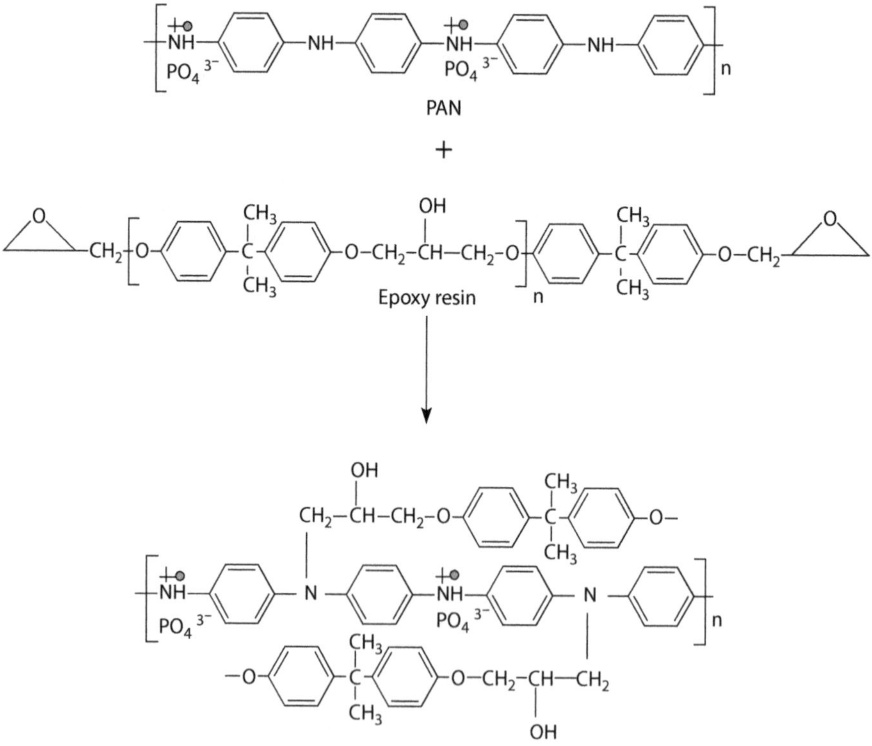

FIGURE 4.16 Scheme of epoxy resin curing by PAN. (From Siva, T., Kamaraj, K., Sathiyanarayanan, S., *Prog. Org. Coat.*, 77, 1095–1103, 2014.)

is a fluctuation in the measured coating resistance values after the initial decrease, indicating the onset of corrosion and immediate passivation at the corroded site as a result of the redox property of the PAN entrapped in the epoxy matrix. Scanning vibrating electrode technique (SVET) provides useful information related with the self-healing capability of the epoxy-PAN composite coating system. As Figure 4.17 shows, SVET indicates a suppression of the initial anodic current flow at the defect area. At the end of the corrosion evaluation test (i.e., after 30 days), the coating resistance was equal to the initial value ($4.7 \times 10^7 \Omega$ cm^2), which is higher than that of epoxy coating cured by polyamide and one order above that of the PAN-free epoxy coating, indicating the better performance of PAN-cured epoxy coating. The presence of PAN entrapped in the cross-linked epoxy matrix helps in improving the corrosion protection performance of the coating system compared with that of the PAN-free epoxy coating. Ennoblement of the open circuit potential and an increase in coating resistance suggest passivation of the steel surface achieved by PAN because of its redox activity.

Copolymers were also used as pigments. Yao et al. (2009) synthesized aniline/p-phenylenediamine copolymer, poly(AN-co-p-DPA), by chemical oxidative polymerization, which has end capped —NH$_2$ structure. The poly(AN-co-p-PDA) with

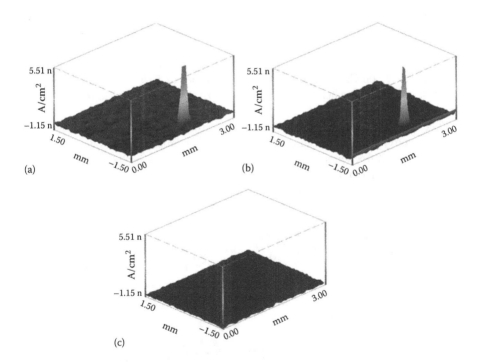

FIGURE 4.17   (**See color insert.**) (a) Current distribution map for PAN-cured epoxy-coated steel in 3 wt% NaCl solution at initial immersion. (b) Current distribution map for PAN-cured epoxy-coated steel immersed in 3 wt% NaCl after 30 min. (c) Current distribution map for PAN-cured epoxy-coated steel in 3 wt% NaCl after 1 h immersion. (From Siva, T., Kamaraj, K., Sathiyanarayanan, S., *Prog. Org. Coat.*, *77*, 1095–1103, 2014.)

end-capped amino groups is synthesized by chemical oxidative polymerization. It was demonstrated not only that the copolymer cures epoxy resin, but also that epoxy resin coating cured with the copolymer can offer good corrosion protection for CS. Improvement of the dispersibility of the P(AN-co-*p*-PDA) copolymers resulted in the best anticorrosive performance. This could be achieved by using chemical oxidative polymerization in the presence of nonionic surfactant nonylphenol ethoxylates (TX-8) with a hydrophilic lipophilic balance (HLB) value of 10–11. The P(AN-co-*p*-PDA)-F copolymer with TX-8 exhibits better dispersion stability in toluene than that without TX-8. Various amounts of P(AN-co-*p*-PDA) were dispersed in *n*-butanol/toluene, while epoxy resin with polyamide was added under stirring conditions (Fu et al., 2013). The P(AN-co-*p*-PDA) synthesized by a traditional oxidative polymerization method displayed wide distribution of particle size, and the maximum of the particle diameter reached to 3 μm (Yao et al., 2009). This was associated with the high polymerization rate of PAN and the aggregation of the primary particles. However, P(AN-co-*p*-PDA)-F displayed a narrow distribution of particle size, and the average particle diameter was reduced to about 0.5 μm as the field emission scanning electron microscopy (FE-SEM) micrographs of Figure 4.18 show. This may be ascribed to the hydrogen bonds between the O groups of TX-8 and the NH groups of P(AN-co-*p*-PDA), which leads to the surface of primary P(AN-co-*p*-PDA) particles being wrapped by the TX-8 and secondary aggregation of P(AN-co-*p*-PDA) particles being restricted. As a consequence, the smaller particle size in P(AN-co-*p*-PDA)-F is obtained.

Polarization and EIS measurements showed that the P(AN-co-*p*-PDA)-epoxy coating applied on CS has good protective properties in 5 wt% NaCl and exhibited the best performance when 7 wt% of P(AN-co-*p*-PDA) was contained in the coating (Figure 4.19).

These results have shown that the anticorrosion property of the P(AN-co-*p*-PDA)-F epoxy coating is caused by both electrochemical protection and barrier effects. The electrochemical protection is caused by the rise in the corrosion potential and the formation of a passive layer on the surface of CS, while paint film prevents penetration of oxygen and H_2O.

FIGURE 4.18   FE-SEM images of (a) P(AN-co-*p*-PDA)-F and (b) P(AN-co-*p*-PDA). (From Fu, P., Li, H., Sun, J., Yi, Z., Wang, G.-C., *Prog. Org. Coat.*, *76*, 589–595, 2013.)

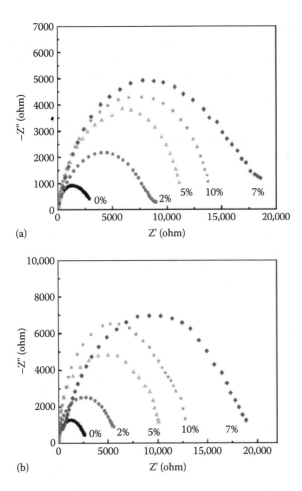

FIGURE 4.19 (See color insert.) Nyquist plots of P(AN-co-*p*-PDA)-F epoxy coating applied on CS with various P(AN-co-*p*-PDA)-F amounts in (a) 5 wt% NaCl and (b) 0.1 M HCl aqueous media for 720 h. (From Fu, P., Li, H., Sun, J., Yi, Z., Wang, G.-C., *Prog. Org. Coat.*, *76*, 589–595, 2013.)

4.3.3 COMPOSITES AND NCS

CP-based composite systems were considered in the corrosion protection with promising properties. Composites, originating from proper combinations of two or more nanosized species by using a suitable technique, result in materials of novel physicochemical properties, different from those of the counterparts. They combine the functional properties of diverse materials imparting improved protective performance. In view of the recent development of nanoscience and nanotechnology, NCs attracted special interest. A number of different metal, metalloid, metal oxide particles or nanoparticles as well as carbon nanomaterials (carbon nanotubes and graphene) can be encapsulated into PPy and PAN to produce composite materials (Ciric-Marjanovic, 2013b).

Using conventional polymers as one component of NCs ensures good process-ability, a property that is necessary for the full exploitation of their potential applications in diverse areas. Other components blended or mixed with the polymer, which is in the solution or in melt form, are often inorganic nanoparticles. The resultant hybrid materials are defined as "polymeric nanocomposites" (Gangopadhyay & De, 2000). A striking difference between conventional NCs and CP-based NCs is that the polymer is not the component that provides flexibility and improves the process-ability of the hybrid material, but the inorganic nanoparticles may provide a degree of processability to the resultant NC. Therefore, conventional blending or mixing techniques to form CP-based NCs with foreign materials cannot be simply used. Several synthesis techniques have been developed and optimized to form composites of CPs with inorganic materials being either as nanoparticles or as nanostructured materials. Inorganic nanoparticles are incorporated into the host π-conjugated system of CPs by using either a chemical or electrochemical path (Gangopadhyay & De, 2000). The anticorrosion properties of composite materials depends on many factors, including the chemical nature and morphology of components, their loadings, distribution, and fabrication technique.

4.3.3.1 Metal Particles

Coatings based on a PAN matrix filled with Zn and Zn nanoparticles were obtained on Fe by solution mixing method (Olad et al., 2011). The presence of Zn nanoparticles in PAN improved its conductivity and anticorrosion properties in comparison with the microsized particles of Zn in PAN composite coatings. In situ production of Zn nanoparticles followed by chemical polymerization of AN, in the presence of Zn nanoparticles, led to PAN–Zn nanocomposite films (Olad & Rasouli, 2010). The PAN–Zn NC was then applied to Fe coupons by the solution casting method. The electrical conductivity of the nanocomposite was found to depend on the Zn loading. The maximum conductivity of the PAN–Zn NC coating was obtained at a certain Zn content. Zn protects iron cathodically by sacrificial protection. The corrosion products of Zn can fill the pores in the PAN resulting in more barrier protection along with the added electrochemical activity of PAN. The iron coupons coated by PAN–Zn NC showed nobler open-circuit potential and lower corrosion current values as compared to iron coupons coated with simple PAN coating. In another study, comparison of the electrical conductivity and anti-corrosion property of PAN–Zn NC coatings was reported with respect to PAN–Zn composite coatings containing Zn micosized particles (Olad et al., 2011). It was shown that the Zn nanoparticles incorporated into the PAN matrix increase the electrical conductivity and improve the protection properties of the PAN–Zn NC coating against iron corrosion in comparison to the Zn microsized particles. The synergistic effect of Zn nanoparticles with PAN was evaluated in acidic chloride media by corrosion tests like Tafel analysis and OCP monitoring. Applying epoxy resin as an additive to the optimized formulation of PAN–Zn NC showed that an optimum range of 3–7 wt% for the epoxy content in the PAN–epoxy–Zn nanocomposite results in coatings that exhibit the best anticorrosion performance (Olad et al., 2012).

4.3.3.2 Metal Oxides

Composite coatings containing ZnO and PAN as nanoadditive dispersions were prepared with poly(vinyl acetate) (PVAc) as the major matrix. The steel plates dip-coated with these formulations were tested for corrosion protection by immersion in saline water over long periods (Patil & Radhakrishnan, 2006). The PAN–ZnO composite coating, containing nanoparticulate ZnO in PVAc matrix, shows significant improvement in corrosion prevention of steel in saline water. Three mechanisms are considered to work synergistically in this case: improvement of barrier properties, formation of p–n junction preventing ease charge transport, and self-healing property of PAN in cases of scratch or scribble. Moreover, these coatings exhibit good gloss and shiny surfaces.

The chemical oxidation of AN monomer with APS was used in the presence of CSA for the preparation of PAN NCs containing ZnO nanorods (Figure 4.20) (Mostafaei & Nasirpouri, 2013, 2014; Mostafaei & Zolriasatein, 2012). The electrical conductivity of the PAN–ZnO NC system was lower than that of pure PAN and ZnO and decreases by increasing the ZnO nanorods. Thermogravimetric analysis (TGA) results showed that the decomposition of the NC was less than that of pure PAN, indicating its potential application in anticorrosive paints. The doping effect of ZnO nanorods content was observed in PAN–ZnO NCs. X-ray diffraction and FTIR results showed that the interaction between PAN and ZnO nanorods is based on the formation of hydrogen bonding and electrostatic interaction between CSA-capped ZnO nanorods and PAN.

PAN–TiO$_2$ composites (PTCs) have been prepared by chemical oxidation of AN and TiO$_2$ by APS in phosphoric acid medium. The PTC was formed on steel by using acrylic resin. The resistance of the coating PTC was evaluated in 3 wt% NaCl

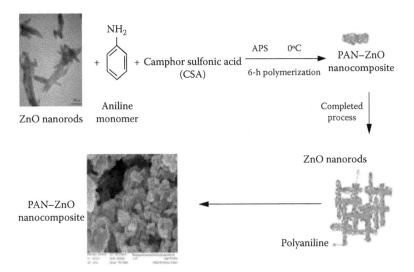

FIGURE 4.20 (**See color insert.**) Preparation steps of PAN–ZnO NC. (From Mostafaei, A., Zolriasatein, A., *Prog. Nat. Sci.*, *22*, 273–280, 2012.)

solution after 60 days by EIS and the salt spray fog test (Sathiyanarayanan et al., 2007a,b). The results indicated that PTC is able to offer higher protection to steel from corrosion in comparison with conventional TiO_2-pigment-containing coating owing to the passivation ability of PAN. Moreover, evaluation results show that PTC exhibits much higher resistance in comparison with that of simple PAN-containing.

PAN-TiO_2 NCs with various dopant percentages of TiO_2 were synthesized at room temperature using a chemical oxidative method (Deivanayaki et al., 2013). Characterization of the PAN-TiO_2 NC showed that the Ti-PAN is formed with an alignment structure of TiO_2 particles. XRD patterns revealed that, as the TiO_2 percentage was increased, the amorphous nature disappeared and the composites became more strongly oriented along the 110 direction. SEM studies revealed the formation of uniform granular morphology with average grain size of 200 nm for PAN-TiO_2 NC at 50 wt% in TiO_2 loading.

Among various inorganic particles, SiO_2 nanoparticles have attracted interest because of their excellent reinforcing properties of polymers (Bhandari et al., 2012). SiO_2 particles were utilized as fillers in metal anticorrosion coatings and found to improve the processibiliy of PAN. Hydrophobic PAN-SiO_2 (HPSC) composites synthesized chemically and deposited on mild steel exhibit improved protective efficiency as compared with PAN. This was explained by considering that the HPSC results in the formation of a passive oxide film on mild steel and simultaneously prevents the chloride ion penetration into the coating owing to the hydrophobic character of the composite. The HPSC coating acts as a barrier between the metal and the corrosive environment. Moreover, the presence of SiO_2 nanoparticles leads to the reinforcement of PAN and, hence, diminution of its degradation (Bhandari et al., 2012). The synergistic effect of the PAN redox catalytic ability and hydrophobic properties leading to advanced anticorrosion coatings was also observed in the case of composites prepared from fluoro-substituded PAN incorporated with silsesquioxane spheres (Weng et al., 2012). Casting of the as-prepared composite material onto a cold-rolled steel (CRS) electrode results in a coated CRS with enhanced corrosion resistance in saline conditions.

The PAN/ferrite/alkyd NC coatings were found to show a far superior corrosion resistance performance compared with that of a pure PAN/alkyd system (Alam et al., 2008). The PAN/ferrite/alkyd was prepared by mixing the appropriate amount of PAN/ferrite with 10 wt% alkyd solution in xylene to obtain different loadings of the NC, varying from 0.5 to 1.5 wt% and applied by brush on mild steel strips. The protection of mild steel strips in chloride-containing acid and neutral environments is ascribed to the high resistance to corrosive ions provided by the dense, nonporous, continuous network-like structure of the PAN/ferrite/alkyd coating. The presence of ferrite particles maintains PAN in its doped state (i.e., prevents the ES to EB transition of PAN) and metal dissolution. Moreover, the small pore size and uniform dispersion of the PAN/ferrite/alkyd NC in (Figure 4.21) favor the formation of a well-adhered, dense structure that impedes the penetration of the corrosive ions through to the metal substrate and inhibit the mild steel from the attack of the corrosive species.

PPy and NCs of PPy with ZnO (PPy–ZnO) were electrodeposited (Hosseini et al., 2011a). The anticorrosion performance of coatings in 3.5 wt% NaCl solution was

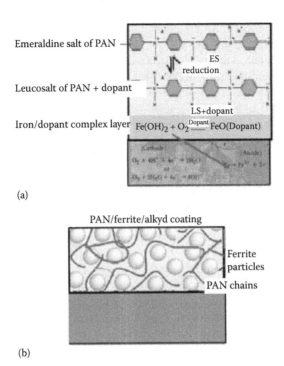

Emeraldine salt of PAN

ES reduction

Leucosalt of PAN + dopant

LS+dopant

Iron/dopant complex layer $Fe(OH)_2 + O_2 \xrightarrow{Dopant} FeO(Dopant)$

(a)

PAN/ferrite/alkyd coating

Ferrite particles

PAN chains

(b)

FIGURE 4.21 (See color insert.) (a) Mechanism of corrosion protection of PAN/alkyd and PAN/ferrite/alkyd coatings. (b) Schematic representation of the structure of PAN/ferrite/alkyd coating indicating the uniform dispersion of the PAN/ferrite nanocomposite in alkyd. (From Alam, J., Riaz, U., Ashraf, S.M., Ahmad, S., *J. Coat. Technol. Res.*, 5, 123–128, 2008.)

monitored by Tafel and EIS analysis. The PPy–ZnO NC coating showed improved corrosion resistance in comparison with the PPy coating. EIS measurements illustrated that the best corrosion protection properties for the investigated coatings in this work are obtained when ZnO nanorods were incorporated in PPy coating. Incorporation of nanosized ZnO (10 wt% relative to PPy) into the PPy matrix resulted in an increase in corrosion resistance of PPy coating. The improved protection performance of PPy–ZnO NC coating over the PPy coating was attributed to the NC morphology in which the particle size and specific surface area are modified with the incorporation of nanorods.

TiO$_2$ nanoparticles, poly(N-methyl pyrrole), coating was electrodeposited on steel surfaces in the presence of dodecyl benzene sulphonic acid and oxalic acid (Mahmoudian et al., 2011a). The presence of TiO$_2$ nanoparticles in the poly(N-methyl pyrrole) coating improved the interactive surface area of poly(N-methyl pyrrole) with the ions involved in the corrosion reaction. This resulted in the decreased water uptake by the composite coating as well as the increase in the pore resistance of the resultant coating.

The performance of PPy-TiO$_2$ NC improved if Sn-doped TiO$_2$ nanoparticles were used in the synthesis of the coating material. PPy/Sn-doped TiO$_2$ NCs were synthesized by chemical polymerization of Py monomer in the presence of Sn-doped TiO_2 particles prepared by the sol–gel method with titanium tetraisopropoxide $(Ti(OCH(CH_3)_2]_4$. The synthesized PPy/Sn-doped TiO_2 NCs powder was combined with epoxy resin and a polyamide hardener, where the PPy/Sn-doped TiO_2 content in the incorporated epoxy-polyamide coating was 1 wt%. Transmission electron microscopy (TEM) and FE-SEM results showed that the Sn-doped TiO_2 NCs had a nucleus effect and caused a homogenous PPy core-shell type of morphology, leading to the coverage of Sn-doped TiO_2 NCs by PPy deposit. The XRD result showed that the crystalline size of PPy/Sn-doped TiO_2 NCs was smaller than that of PPy/TiO_2 NCs. Due to the decrease of the crystalline size of the PPy/Sn-doped TiO_2 NC its barrier effect is improved when was used as anticorrosive pigment in epoxy primer on steel substrate (Mahmoudian et al., 2011). In addition to this factor, a suggested protection mechanism was based, first, on the increased area of PPy-TiO_2 nanoparticles interacting with ions produced during the corrosion reaction of steel in 3.5 wt% NaCl solution, and, second, on the effect of Sn doping on the charge transfer processes involved in the formation of rust and dissolution of steel. Figure 4.22 illustrates a schematic diagram for the energy levels of the materials in contact within the PPy/Sn-doped TiO_2 NC by which possibel electron transfer processes can be understood. Doped PPy exhibits p-type semiconductivity in contrast to the n-type TiO_2 and SnO_2 semiconductors. By the coupling of TiO_2 and Sn nanoparticles, the SnO_2 acts as sink of electrons due to the positive shift of its conduction band (CB) with respect to that of TiO_2. The tendency for the formation of a compact passive oxide layer on the steel surface by PPy can be explained by considering the different energy levels of PPy and Fe. Reduction of PPy promotes the catalytic formation of a stable oxide layer enriched in ferric ions via the transformation of a water-soluble oxide enriched in ferrous ions.

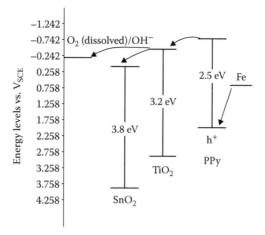

FIGURE 4.22 Schematic diagram of the energy levels of the PPy/Sn-doped TiO_2 NC applied onto a steel surface. (From Mahmoudian, M.R., Basirun, W.J., Alias, Y., Ebadi, M., *Appl. Surf. Sci.*, *257*, 8317–8325, 2011.)

Along the same lines and by an analogous mechanism, PPy/Ni-doped TiO_2 nanoparticles were found to offer also improved protection to steel surfaces as compared with the PPy-TiO_2 coating (Mahmoudian et al., 2011b). The EIS results support a better protection performance. The principal reason for better performance was the increase in the surface area caused by the presence of TiO_2 nanoparticles that resulted in increased interaction between the PPy and the ions involved in the corrosion reaction.

Electrodeposition of poly(M-methyl pyrrole) (PMPy) on steel in the presence of TiO_2 nanoparticles in a mixture of dodecyl benzene sulfonic acid (DBSA) with oxalic acid was proved to be another efficient process for the synthesis of PMPy-TiO_2 nanocomposites. The TiO_2 nanoparticles induce notable changes in the morphology of the PMPy films resulting in a decrease of the amount of the water uptake in 3.5 wt% NaCl solution. This factor along with the increased ability of the PMPy-TiO_2 NC to interact with the steel substrate improves the anticorrosion property of the PMPy-TiO_2 NC in comparison with that of PMPy films (Mahmoudian et al., 2011a).

4.3.3.3   Carbon-Based NCs

Composites of PAN with multi-walled carbon nanotubes (MWCNT) used as paints on low-carbon steel were found to have promising anticorrosion properties (Deshpande et al., 2013). For example, the presence of PAN-MWCNT-based paints decreases significantly the corrosion rate of the steel in 3.5 wt% NaCl aqueous solution. MWCNT served also as dopants in poly(3,4-ethylenedioxythiophene) (PEDOT) nanospheres (NSP) for the preparation of the PEDOT-NSP/MWCNT composite material (Prabakar & Pyo, 2012). Due to the presence of the negatively charged MWCNT, the intrinsic anion-exchange selectivity of PEDOT changes to a rather cation-exchange selectivity. The PEDOT-NSP/MWCNT was synthesized using microemulsion polymerization and investigated as a corrosion inhibitor for aluminum in $LiPF_6$. The anticorrosion effectiveness of PEDOT-NSP/MWCNT was attributed to a synergistic effect that involves the cation exchange PEDOT and the anion-repulsive pristine MWCNT surface. Due to this cooperative action, the transport of the PF_6^- anions towards the aluminum surface is blocked and hence pitting corrosion induced by PF_6^- anions is prevented.

PPy/carboxylic acid-functionalized single-walled carbon nanotubes (CNT-CA), as well as PPy/poly-aminobenzene sulfonic acid-functionalized single-walled carbon nanotubes (CNT-PABS) were electrochemically prepared from oxalate solutions (Ionita & Pruna, 2011). In this galvanostatic synthesis, the CNT-CA and CNT-PABS served as dopants in the resulting PPy/CNT-CA and PPy-CNT-PABS composite layers on carbon steel (CS) surfaces. Moreover, experimental and computational studies showed that the incorporation of CNT-CA and CNT-PABS induces mechanical reinforcement in PPy. The reinforcement of PPy was found to depend on the CNT loading be optimized at certain content of CNT. The PPy/CNT-CA and PPy/CNT-PABS composite layers on CS examined as protective materials in 3.5 wt% NaCl solution. It was found that CS surfaces coated with the composite coatings showed more noble corrosion potential and lower corrosion current than CS surfaces coated with PPy. The PPy/CNT-PABS was more stable than the PPy/CNT-CA composite layer.

Research activities associated with conductive graphene with a relatively high aspect ratio of ~500 have evoked interest in fabricating graphene/polymer composites for various applications (Ansari et al., 2014; Chang et al., 2012; Luo, Jiang et al., 2013; Luo, Zhu et al., 2013). The lower density and higher aspect ratio of conductive graphene, as compared with that of nonconductive clay platelets, initiated their potential application as advanced gas barrier polymer composite films. Nanocasting was used to develop epoxy/graphene composites as corrosion inhibitors with hydrophobic surfaces (Chang et al., 2014).

Polyaniline/graphene composites (PAGCs) that incorporate aminobenzoyl group-functionalized graphene-like (ABF-G) sheets in PAN for corrosion protection coatings were prepared. The PAGCs were obtained by exfoliating and functionalizing by direct electrophilic substitution reaction with 4-aminobenzoic acid (ABA) in a polyphosphoric acid/P_2O_5 medium. The subsequent chemical oxidation polymerization of the AN monomers with different amounts of ABF-G sheets was conducted using APS as oxidant in 1 M HCl to yield PAGCs. Well-dispersed graphene, with a relatively high aspect ratio compared with clay, in a polymer matrix enhances the gas barrier and is responsible for the highly desirable anticorrosion properties that make PAGCs much more effective than PACCs. As such, PAGCs have an excellent potential to be used as corrosion protection coating materials (Chang et al., 2012). The coatings were shown to effectively protect steel because of the good O_2 and H_2O gas barrier. It was found that polystyrene film with a low graphene loading was superior in reducing the relative O_2 permeability of PS compared with those of the best published gas barrier results for polymer/clay composites (Figure 4.23).

Because it is easy to obtain the graphene precursor, graphite, as it is naturally abundant, and the functionalized graphene can serve as a conductive nanofiller for other polymers (such as epoxy, polyimide, polyurethane), polymer/graphene composites will launch a new era of corrosion protection materials in the future (Chang et al., 2012).

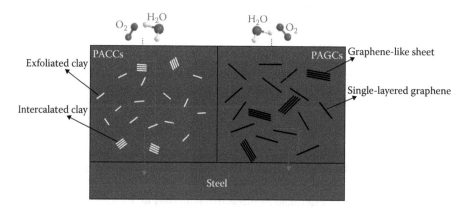

FIGURE 4.23 (See color insert.) Schematic representation of O_2 and H_2O following a tortuous path through PACCs and PAGCs. (From Chang, C.-H., Huang, T.-C., Peng, C.-W., Yeh, T.-C., Lu, H.-I., Hung, W.-I., Weng, C.-J., Yang, T.-I., Yeh, J.-M., *Carbon, 50*, 5044–5051, 2012.)

4.3.3.4   Layered Materials

Composites and NCs of CPs can be prepared by the inclusion of layered materials. The plate-like shape of these materials lengthens the path of corrosive ions toward the substrate, resulting in delay in the onset of corrosion process. The flake structure can reduce the permeability of the solute by a factor of 10 after their inclusion in the coating (Yang et al., 2004). CP/montmorillonite (MMT) clay composites and NCs have been used for the corrosion protection of metals and their alloys (Akbarinezhad et al., 2011; Hosseini et al., 2011b; Shabani-Nooshabadi et al., 2011). PAN/MMT NCs were prepared by blending PAN and MMT in epoxy resin (Hosseini et al., 2011b). The specific morphology of the NC resulted in the improved corrosion protection of Al 5000.

PAN/clay NC was prepared by a chemical oxidative polymerization method (Akbarinezhad et al., 2011). This NC was then incorporated in a zinc-rich ethyl silicate resin. EIS measurements and OCP monitoring were carried out in 3.5 wt% NaCl solution for the corrosion evaluation of coated carbon steel panels with the PAN/clay NC and simple PAN coating for 120 days of immersion. The barrier protection and the passivation of the substrate due to the presence of the composite pigment resulted in the improved corrosion resistance of the coating containing PAN/clay NC in comparison with the pristine coating. Along similar lines, PAN/MMT NC coatings were prepared on aluminum alloy (AA 3004) by the galvanostatic method (Shabani-Nooshabadi et al., 2011). The corrosion performance properties of these coatings were monitored in 3.5 wt% NaCl solution by EIS and potentiodynamic polarization measurements. With the NC on the surface of AA 3004, the corrosion current was found to decrease. The anticorrosion performance of the PAN/MMT NC was attributed to the barrier nature of MMT and redox catalytic nature of PAN.

PAN/precipitated calcium carbonate (PCC) composite material was prepared by a convenient route using naturally occurring calcite for use as anticorrosive coatings on metal surfaces. The PCC produced by this approach is present in its vaterite form. The coating is prepared by dispersing the composite in a mixture of xylene and alkyd resin. Stabilized vaterite nanoparticles with an average crystallite size of 26 nm can be synthesized by PAN as a template. As was confirmed by FE-SEM images, PAN chains are filled by PCC nanoparticles to form PAN/PCC spheres with an average diameter of 3–4 μm. The PAN/PCC nanocomposites exhibit sufficient electrical conductivity and can be used as anticorrosive surface coatings on metal surfaces and to produce vaterite nanoparticles. FTIR spectra indicated different PAN:vaterite ratios. The PAN/vaterite composites and a PAN/calcite composite were mixed with either alkyd resin or xylene to prepare anticorrosive coatings on mild steel of approximately same thickness. Corrosion evaluation measurements indicated that the PAN/vaterite-alkyd resin composites show better protective properties for mild steel than the PAN/calcite-alkyl resin coating (Cathuranga et al., 2014).

PPy/Al flake composite coatings have been found to protect aluminum alloy (AA 2024) against corrosion (Yan et al., 2010). The presence of PPy in intimate contact with aluminum flakes activates them toward anodic activity, resulting in galvanic coupling between the composite and the substrate.

4.3.4 Nanostructured CPs

Nanostructured CPs can enhance the anticorrosion performance of CP by improving charge transport rate as well as increasing surface area. Nanostructured materials have been establishing themselves as the modern generation of high-performance materials in many areas, ranging from automotive engineering to bioengineering, owing to a vast array of unique properties. The tiny size of the nanoparticles produces an extraordinarily high surface energy, an increased number of surface atoms that exhibit enhanced compactness, and physico-mechanical and physico-chemical resistance compared with common microparticles. Advancements in modern engineering and technology have hastened the development of high-performance, corrosion-resistant coatings that have a broad spectrum of effectiveness under a wider range of hostile environments.

For example, the anticorrosive performance of a water-based epoxy coating on steel substrate was improved drastically by adding only 0.02 wt% nano-PAN (Bagzerzadeh et al., 2010). PAN nanoparticles at low loading in PAN-based alkyd coating exhibited also good protective performance against the corrosion of mild steel, although in comparison with the efficiency observed by the poly(1-naphylamine)-based alkyl coating, it was worse (Riaz et al., 2009).

Comparison between the effect of PAN nanotubes of a narrow diameter and PAN nanofibers in polyvilyl butyral coating on the protective performance of mild steel showed that the resistance of the coating containing PAN nanotubes is three times higher than the coating containing PAN nanofibers after 30 days of immersion in 3.5 wt% NaCl solution. This effect was assigned to the higher surface area of the nanotubes compared with nanofibers with the same mass. The nanotubular PAN has a greater ability to interact with the ions liberated during the corrosion of the steel and increase the rate of the cathodic reduction of oxygen on the PAN surface (Mahmoudian et al., 2012). PAN nanofibers synthesized by interfacial polymerization were found to have more excellent protection against the corrosion of CS in 5 wt% NaCl solution as compared with the aggregated PAN (Yao et al., 2008).

Comparative studies have shown that the nanostructured and morphology of PAN could influence its anticorrosion efficiency. PAN with different nanostructures was synthesized by (1) conventional polymerization, (2) interfacial polymerization, and (3) direct mixed reaction methods in sulfuric acid solutions (Yang et al., 2010). SEM images of Figure 4.24 (left-hand side) show the different morphologies of PAN prepared by the three methods. The nanostructured PAN products were applied to a clean mild steel surface as a 10% suspension in NMP, respectively. The PAN coating was allowed to dry in air at room temperature for 72 h, and its morphology was as Figure 4.24 (right-hand side) illustrates. This comparative study shows the important role of preparation method. Conventional polymerization results in irregularly shaped agglomerates containing short PAN nanorods and particulates. The PAN prepared by interfacial polymerization exhibits fibrillar morphology with nanofibers of diameters, $d = 80–150$ nm and lengths varying from 500 nm to several micrometers. The PAN nanofibers synthesized by direct mixing have diameters in the range 60–100 nm and several micrometer length.

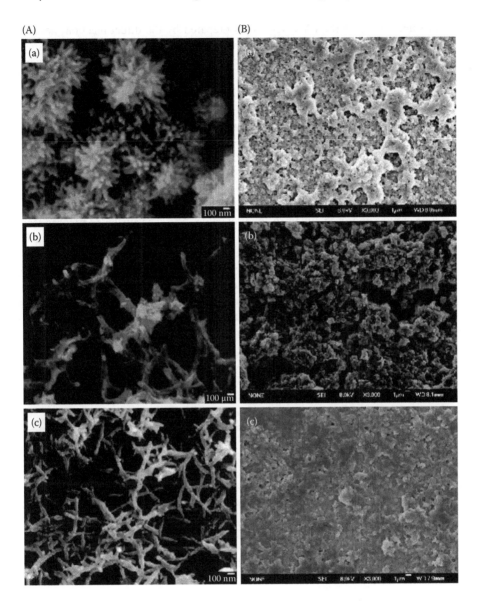

FIGURE 4.24 SEM micrographs of as-prepared (column A) and the coatings (column B) applied to mild steel of the nanostructured PAN: (a) conventional polymerization, (b) interfacial polymerization, and (c) direct mixed method. (From Yang, X.G., Li, B., Wang, H.Z., Hou, B.R., *Prog. Org. Coat.*, *69*, 267–271, 2010.)

These nanofibers form interconnected networks resulting in a uniform morphology. Apparently stirring in the direct mixed reaction enhanced the uniform distribution and diffusion of monomer AN that facilitated the growth of PAN nanofibers into larger sized fibers by preventing the formation of larger sized particles and agglomerates.

The uniformity of the PAN nanofibers prepared by the direct reaction method depends on the inorganic acid used for the PAN doping (Ge et al., 2012). Using H_2SO_4, H_3PO_4, HNO_3, and HCl showed that uniformity, which impacts also anticorrosion performance, varied in the order $H_3PO_4 > H_2SO_4 > HNO_3 > HCl$.

The PAN was applied to the mild steel surface as a suspension in NMP at a thickness of 5 ± 0.1 μm. The anticorrosion performance of PAN coatings was tested by potentiodynamic polarization measurements. The corrosion potential, E_{corr}, the corrosion current density, I_{corr}, and the protective efficiency estimated from the I_{corr} obtained by Tafel analysis are summarized in Table 4.3, along with the morphology of the as-synthesized PAN and the features of its coatings on mild steel observed in SEM images.

As seen in Table 4.3, coatings of PAN nanofibers synthesized by a direct mixed reaction exhibit improved anticorrosion protection. This was attributed to three features of this PAN-based coating: (1) more compact and uniform structure, (2) nano fibrillar ordered structure giving rise to an increased contact with the substrate that promotes the formation of compact passive layer on the mild steel surface and a negative shift of the cathodic reaction, and (3) perhaps crystalline structure. XRD characterization showed that the PAN prepared by the direct mixed reaction exhibits a higher degree of crystallinity as compared with PAN obtained by conventional and interfacial polymerization. Therefore, besides the more compact structure obtained in the case of a nanofiber network, increasing the order in nanostructured PAN, which results in a higher conductivity, may also contribute to an improved protective performance. The effect of crystalline CPs on their anticorrosion properties is an open question and remains to be investigated.

Table 4.3

Effect of the Morphology of Nanostructured PAN Synthesized by Different Methods on the Protective Efficiency of PAN Coatings Applied on Mild Steel

Polymerization Method	Morphology	Coating Features	E_{corr} (mV$_{SCE}$)	I_{corr} (μA cm^{-1})	Inhibition Efficiency (%)
Conventional	Irregularly shaped agglomerates containing nanorods and particulates	Porous	-550.10 ± 3.69	8.57 ± 0.06	21.43 ± 0.19
Interfacial	Nanofibers ($d = 80$–150 nm)	Cracks	-613.44 ± 5.77	6.27 ± 0.05	42.57 ± 0.34
Direct mixed reaction	Interconnected nanofiber networks ($d = 60$–100 nm)	Uniform and compact	-654.20 ± 4.87	3.48 ± 0.02	68.08 ± 0.29

Source: Yang, X.G., Li, B., Wang, H.Z., Hou, B.R., *Prog. Org. Coat.*, 69, 267–271, 2010.

REFERENCES

Adamczyk, L., & Kulesza, P. J. (2011). Fabrication of composite coatings of 4-(pyrrole-1-yl) benzoate-modified poly-3,4-ethylenedioxythiophene with phosphomolybdate and their application in corrosion protection. *Electrochim. Acta*, *56*, 3649–3655.

Adamczyk, L., Giza, K., & Dudek, A. (2014). Electrochemical preparation of composite coatings of 3,4-etylenodioxythiophene (EDOT) and 4-(pyrrole-1-yl) benzoic acid (PyBA) with heteropolyanions. *Mater. Chem. Phys.*, *144*, 418–424.

Aeiyach, S., Bazzaouri, E. A., & Lacaze, P. C. (1997). Eelctropolymerization of thiophene on oxidizable metals in organic media. *J. Electroanal. Chem.*, *434*, 153–162.

Akbarinezhad, E., Ebrahimi, M., Sharif, F., Attar, M. M., & Faridi, H. R. (2011). Synthesis and evaluating corrosion protection effects of emeraldine base PAni/clay nanocomposite as a barrier pigment in zinc-rich ethyl silicate primer. *Prog. Org. Coat.*, *70*(1), 39–44.

Alam, J., Riaz, U., Ashraf, S. M., & Ahmad, S. (2008). Corrosion protective performance of nano polyaniline/ferrite dispersed alkyl coatings. *J. Coat. Technol. Res.*, *5*, 123–128.

Anderson, N., Irvin, D. J., Webber, C., Stenger-Smith, J. D., & Zarras, P. (2002). Scale-up and corrosion inhibition of poly(bis-dialkylamino)phenylene vinylene)s. *PMSE Prepr.*, *223*, 57.

Ansari, M. O., Khan, M. M., Ansari, S. A., Amal, I., Lee, J. G., & Cho, M. H. (2014). Enhanced thermoelectric performance and ammonia sensing properties of sulfonated polyaniline/graphene thin films. *Mater. Lett.*, *114*, 159–162.

Arefinia, R., Shojaei, A., Shariatpanahi, H., & Neshati, J. (2012). Anticorrosion properties of smart coating based on polyaniline nanoparticles/epoxy–ester system. *Prog. Org. Coat.*, *75*(4), 502–508.

Armelin, E., Ocampo, C., Liesa, F., Iribarren, J., Ramis, X., & Aleman, C. (2007a). Study of epoxy and alkyd coatings modified with emeraldine base form of polyaniline. *Prog. Org. Coat.*, *58*, 316–322.

Armelin, E., Oliver, R., Liesa, F., Iribarren, J., Estrany, F., & Aleman, C. (2007b). Marine paint formulations: Conducting polymers as anticorrosive additives. *Prog. Org. Coat.*, *59*, 46–52.

Armelin, E., Pla, R., Liesa, F., Ramis, X., Iribarren, J. I., & Aleman, C. (2008). Corrosion protection with polyaniline and polypyrrole as anticorrosive additives for epoxy paint. *Corros. Sci.*, *50*(3), 721–728.

Armelin, E., Meneguzzi, A., Ferreira, C. A., & Alemán, C. (2009). Polyaniline, polypyrrole and poly(3,4-ethylenedioxythiophene) as additives of organic coatings to prevent corrosion. *Surf. Coat. Technol.*, *203*, 3763–3769.

Ashraf, S. A., Chen, F., Too, C. O., & Wallace, G. G. (1996). Bulk electropolymerization of alkylpyrroles. *Polymer*, *37*, 2811–2819.

Bagzerzadeh, M. R., Mahdavian, F., Ghasemi, M., Shariatpanahi, H., & Faridi, H. R. (2010). Using nanoemeraldine salt–polyaniline for preparation of a new anticorrosive water-based epoxy coating. *Prog. Org. Coat.*, *68*, 319–322.

Balaskas, A. C., Kartsonakis, I. A., Kordas, G., Cabral, A. M., & Morais, P. J. (2011). Influence of the doping agent on the corrosion protection properties of polypyrrole grown on aluminium alloy 2024-T3. *Prog. Org. Coat.*, *71*(2), 181–187.

Baldissera, A. F., & Ferreira, C. A. (2012). Coatings based on electronic conducting polymers for corrosion protection of metals. *Prog. Org. Coat.*, *75*, 241–247.

Baldissera, A. F., Freitas, D. B., & Ferreira, C. A. (2010). Electrochemical impedance spectroscopy investigation of chlorinated rubber-based coatings containing polyaniline as anticorrosion agent. *Mater. Corros.*, *61*, 790–801.

Bard, A. J., & Faulkner, L. R. (2000). *Electrochemical Methods: Fundamentals and Applications*. New York: John Wiley & Sons.

Barsch, U., & Beck, F. (1993). Electrodeposition of polythiophene from bisthiophene onto iron. *Synth. Met.*, *55*, 1638–1643.

Bazzaouri, E. A., Aeiyach, S., & Lacaze, P. C. (1996). Electropolymerization of bisthiophene on Pt and Fe electrodes in an aqueous SDS micellar medium. *Synth. Met.*, *83*, 159–165.

Beck, F., & Michaelis, R. (1992). Strongly adherent, smooth coatings of polypyrrole oxalate on iron. *J. Coat. Technol.*, *64*, 59–67.

Beck, F., Michaelis, R., Schloten, F., & Zinger, B. (1994). Filmforming electropolymerization of pyrrole on iron in aqueous oxalic acid. *Electrochim. Acta*, *39*, 229–234.

Bernard, M. C., Hugot-Le Goff, A., Joiret, S., Dinh, N. N., & Toan, N. N. (1999). Polyaniline layer for iron protection in sulfate medium. *J. Electrochem. Soc.*, *146*(3), 995–998.

Bernard, M. C., Joiret, S., Hugot-Le Goff, A., & Phong, P. V. (2001). Protection of iron against corrosion using a polyaniline layer I. Polyaniline electrodeposit. *J. Electrochem. Soc.*, *148*, B12–B16.

Bhadra, J., Madi, N. K., Al-Thani, N. J., & Al-Maadeed, M. A. (2014). Polyaniline/polyvinyl alcohol blends: Effect of sulfonic acid dopants on microstructural, optical, thermal and electrical properties. *Synth. Met.*, *191*, 126–134.

Bhandari, H., Anoop Kumar, S., & Dhawan, S. K., eds. (2012). *Conducting Polymer Nanocomposites for Anticorrosive and Antistatic Applications*. Rejeka, Croatia: INTECHOPEN. COM: INTECH.

Biallozor, S., & Kupniewska, A. (2005). Conducting polymers electrodeposited on active metals. *Synth. Met.*, *155*(3), 443–449.

Breslin, C. B., Fenelon, A., M., & Conroy, K. G. (2005). Surface engineering: Corrosion protection using conducting polymers. *Mater. Des.*, *26*, 233–237.

Çakmakcı, İ., Duran, B., Duran, M., & Bereket, G. (2013). Experimental and theoretical studies on protective properties of poly(pyrrole-co-N-methyl pyrrole) coatings on copper in chloride media. *Corros. Sci.*, *69*, 252–261.

Camalet, J. L., Lacroix, J. C., Aelyach, S., Chane-Ching, K. I., & Lacaze, P. C. (1996). Electrodeposition of protective polyaniline films on mild steel. *J. Electroanal. Chem.*, *416*, 179–182.

Camalet, J. L., Lacroix, J. C., Aeiyach, S., Chane-Ching, K. I., & Lacaze, P. C. (1998a). Electrosynthesis of adherent polyaniline films on iron and mild steel in aqueous oxalic acid medium. *Synth. Met.*, *93*, 133–142.

Camalet, J. L., Lacroix, J. C., Aeiyach, S., & Lacaze, P. C. (1998b). Characterization of polyaniline films electrodeposited on mild steel in aqueous p-toluenesulfonic acid solution. *J. Electroanal. Chem.*, *445*, 117–124.

Chandrasekhar, P. (1999). *Conducting Polymers, Fundamentals and Applications: A Practical Approach*. Norwell, MA: Kluwer Academic Publishers.

Chang, C.-H., Huang, T.-C., Peng, C.-W., Yeh, T.-C., Lu, H.-I., Hung, W.-I., Weng, C.-J., Yang, T.-I., & Yeh, J.-M. (2012). Novel anticorrosion coatings prepared from polyaniline/graphene composites. *Carbon*, *50*(14), 5044–5051.

Chang, K.-C., Hsu, M.-H., Lu, H.-I., Lai, M.-C., Liu, P.-J., Hsu, C.-H., Ji, W.-F., Chuang, T.-L., Wei, Y., Yeh, J.-M., & Liu, W.-R. (2014). Room-temperature cured hydrophobic epoxy/graphene composites as corrosion inhibitor for cold-rolled steel. *Carbon*, *66*, 144–153.

Chathuranga Senarathna, K. G., Mantilaka, M. M. M. G. P. G., Nirmal Peiris, T. A., Pitawala, H. M. T. G. A., Karunaratne, D. G. G. P., & Rajapakse, R. M. G. (2014). Convenient routes to synthesize uncommon vaterite nanoparticles and the nanocomposites of alkyd resin/polyaniline/vaterite: The latter possessing superior anticorrosive performance on mild steel surfaces. *Electrochim. Acta*, *117*, 460–469.

Cho, S. I., & Lee, S. B. (2008). Fast electrochemistry of conductive polymer nanotubes: Synthesis, mechanism and application. *Acc. Chem. Res.*, *41*(6), 699–707.

Chujo, Y. (2010). *Conducting Polymer Synthesis—Methods and Reactions*. Weinheim: Wiley-VCH Verlag GmbH & Co. KGaA.

Ciric-Marjanovic, G. (2013a). Recent advances in polyaniline research: Polymerization mechanisms, structural aspects, properties and applications. *Synth. Met.*, *177*, 1–47.

Ciric-Marjanovic, G. (2013b). Recent advances in polyaniline composites with metals, metalloids and nonmetals. *Synth. Met.*, *170*, 31–56.

DeBerry, D. W. (1985). Modification of the electrochemical and corrosion behavior of stainless steels with an electroactive coating. *J. Electrochem. Soc.*, *132*(5), 1022–1026.

Deivanayaki, S., Ponnuswamy, V., Ashokan, S., Jayamurugan, P., & Mariappan, R. (2013). Synthesis and characterization of TiO2-doped Polyaniline nanocomposites by chemical oxidation method. *Mater. Sci. Semicond. Process.*, *16*(2), 554–559.

Deshpande, P. P., Vathare, S. S., Vagge, S. T., Tomsik, E., & Stejskal, J. (2013). Conducting polyaniline/multi-wall carbon nanotubes composite paints on low carbon steel for corrosion protection: Electrochemical investigations. *Chem. Pap.*, *67*, 1072–1078.

Deshpande, P. P., Jadhav, N. G., Gelling, V. J., & Sazou, D. (2014). Conducting polymers for corrosion protection: A review. *J. Coat. Technol. Res.*, *11*, 473–494.

Dudukcu, M., Udum, Y. A., Ergun, Y., & Koleli, F. (2009). Electrodeposition of poly(4-methylcarbazole-3-carboxylic acid) on steel surfaces and corrosion protection of steel. *J. Appl. Polym. Sci.*, *111*, 1496–1500.

Duran, B., Çakmakcı, İ., & Bereket, G. (2013). Role of supporting electrolyte on the corrosion performance of poly(carbazole) films deposited on stainless steel. *Corros. Sci.*, *77*, 194–201.

Eftekhari, A., ed. (2010). *Nanostructured Conductive Polymers*. Chichester, UK: John Wiley & Sons Ltd.

Elkais, A. R., Gvozdenović, M. M., Jugović, B. Z., & Grgur, B. N. (2013). The influence of thin benzoate-doped polyaniline coatings on corrosion protection of mild steel in different environments. *Prog. Org. Coat.*, *76*(4), 670–676.

Ferreira, C. A., Aeiyach, S., Aaron, J. J., & Lacaze, P. C. (1996). Electrosynthesis of strongly adherent Ppy coatings on iron and mild steel in aqueous media. *Electrochim. Acta*, *41*, 1801–1809.

Flamini, D. O. (2010). Electrodeposition of polypyrrole onto NiTi and the corrosion behaviour of the coated alloy. *Corros. Sci.*, *52*, 229–234.

Flamini, D. O., Saugo, M., & Saidman, S. B. (2014). Electrodeposition of polypyrrole on Nitinol alloy in the presence of inhibitor ions for corrosion protection. *Corros. Sci.*, *81*, 36–44.

Forsgren, A. (2006). *Corrosion Control through Organic Coatings*. Boca Raton, FL: CRC Press Taylor & Francis Group.

Frau, A. F., Pemites, R. B., & Advincula, R. C. (2010). A conjugated polymer network approach to anticorrosion coatings: Poly(vinylcarbazole) electrodeposition. *Ind. Eng. Chem. Res.*, *49*, 9789–9797.

Freund, M. S., & Deore, B. A. (2007). *Self-Doped Conducting Polymers*. New York: Wiley.

Fu, T., Liu, J., Wang, J., & Na, H. (2009). Cure kinetics and conductivity of rigid rod epoxy with polyaniline as a curing agent. *Polym. Compos.*, *30*, 1394–1400.

Fu, P., Li, H., Sun, J., Yi, Z., & Wang, G.-C. (2013). Corrosive inhibition behavior of well-dispersible aniline/p-phenylenediamine copolymers. *Prog. Org. Coat.*, *76*, 589–595.

Gangopadhyay, R., & De, A. (2000). Conducting polymer nanocomposites: A brief overview. *Chem. Mater.*, *12*, 608–622.

Ge, C. Y., Yang, X. G., & Hou, B. R. (2012). Synthesis of polyaniline nanofiber and anticorrosion property of polyaniline-epoxy composite coating for Q235 steel. *J. Coat. Technol. Res.*, *9*, 59–69.

Gonzalez, M. B., & Saidman, S. B. (2011). Electrodeposition of polypyrrole on 316L stainless steel for corrosion prevention. *Corros. Sci.*, *53*(1), 276–282.

Gospodinova, N., & Terlemezyan, L. (1998). Conducting polymers prepared by oxidative polymerization: Polyaniline. *Prog. Polym. Sci.*, *23*, 1443–1484.

Grundmeler, G., & Simoes, A. (2007). Corrosion protection by organic coatings. In: J. Bard, M. Stratmann & G. S. Frankler (Eds.), *Encyclopedia of Electrochemistry*. vol. 4 (pp. 500–566). Weinheim, Germany: Wiley-VCH Verlag GmbH & Co. KGaA.

Guo, Y. P., & Zhou, Y. (2007). Polyaniline nanofibers fabricated by electrochemical polymerization: A mechanistic study. *Eur. Polym. J.*, *43*(6), 2292–2297.

Guo, X. W., Jiang, Y. F., Zhai, C. Q., Lu, C., & Ding, W. J. (2003). Preparation of even polyaniline film on magnesium alloy by pulse potentiostatic method. *Synth. Met.*, *135–136*, 169–170.

Gupta, G., Birbilis, N., Cook, A. B., & Khanna, A. S. (2013). Polyaniline-lignosulfonate/epoxy coating for corrosion protection of AA2024-T3. *Corros. Sci.*, *67*, 256–267.

Gzgur, B. N., Elkais, A. R., Gvozdenovic, M. M., Drmanic, S. Z., Trisovic, T. L., & Jugovic, B. Z. (2015). Corrosion of mild steel with composite coatings using different formulations. *Prog. Org. Coat.*, *79*, 17–24.

Havinga, E. E., Ten Hoeve, W., Meijer, E. W., & Wynberg, H. (1989). Water-soluble self-doped 3-substituted polypyrroles. *Chem. Mater.*, *1*, 650–659.

Herrasti, P., del Rio, A. L., & Recio, J. (2007). Electrodeposition of homogeneous and adherent polypyrrole on copper for corrosion protection. *Electrochim. Acta*, *52*(23), 6496–6501.

Hosseini, M. G., Bagheri, R., & Najjar, R. (2011a). Electropolymerization of polypyrrole and polypyrrole–ZnO nanocomposites on mild steel and its corrosion protection performance. *J. Appl. Polym. Sci.*, *121*(6), 3159–3166.

Hosseini, M. G., Jafari, M., & Najjar, R. (2011b). Effect of polyaniline–montmorillonite nanocomposite powders addition on corrosion performance of epoxy coatings on Al 5000. *Surf. Coat. Technol.*, *206*(2–3), 280–286.

Hulser, P., & Beck, F. (1990). Electrodeposition of polypyrrole layers on aluminium from aqueous electrolytes. *J. Appl. Electrochem.*, *20*, 596–605.

Inzelt, G. (2008). *Conducting Polymers: A New Era in Electrochemistry*. Berlin, Heidelberg: Springer.

Ionita, M., & Pruna, A. (2011). Polypyrrole/carbon nanotube composites: Molecular modeling and experimental investigation as anti–corrosive coating. *Prog. Org. Coat.*, *72*(4), 647–652.

Iribarren, J. I., Liesa, F., Cadena, F., & Bilurbina, L. (2004). Urban and marine corrosion: Comparative behaviour between field and laboratory conditions. *Mater. Corr.*, *55*, 689–694.

Iribarren, J. I., Cadena, F., & Liesa, F. (2005). Corrosion protection of carbon steel with thermoplastic coatings and alkyd resins containing polyaniline as conductive polymer. *Prog. Org. Coat.*, *52*, 151–160.

Irvin, D. J., Anderson, N., Webber, C., Fallis, S., & Zarras, P. (2002). New synthetic routes to poly(bis-(dialkylamino)phenylene vinylene)s (BAM-PPV). *PMSE Prepr.*, *223*, 67.

Jiang, Y. F., Guo, X. W., Wei, Y. H., Zhai, C. Q., & Ding, W. J. (2003). Corrosion protection of polypyrrole electrodeposited on AZ91 magnesium alloys in alkaline solutions. *Synth. Met.*, *139*(2), 335–339.

Karpakam, V., Kamaraj, K., Sathiyanarayanan, S., Venkatachari, G., & Ramu, S. (2011). Electrosynthesis of polyaniline-molybdate coating on steel and its corrosion protection performance. *Electrochim. Acta*, *56*(5), 2165–2173.

Khan, M. I., Chaudhry, A. U., Hashim, S., Zahoor, M. K., & Igbal, M. Z. (2010). Recent developments in intrinsically conductive polymer coatings for corrosion protection. *Chem. Eng. Res. Bull.*, *14*, 73–86.

Kinlen, P. J., Menon, V., & Ding, Y. W. (1999). A mechanistic investigation of polyaniline corrosion protection using the scanning reference electrode technique. *J. Electrochem. Soc.*, *146*(10), 3690–3695.

Kinlen, P. J., Ding, Y., & Silverman, D. C. (2002). Corrosion protection of mild steel using sulfonic and phosphonic acid-doped polyanilines. *Corrosion*, *58*(6), 490–497.

Koul, S., Dhawan, S. K., & Chandra, R. (2001). Compensated sulfonated polyaniline—Correlation of processibility and crystalline structure. *Synth. Met.*, *124*, 295–299.

Kousik, G., Pitchumani, S., & Renganathan, N. G. (2001). Electrochemical characterization of polythiophene-coated steel. *Prog. Org. Coat.*, *43*(4), 286–291.

Kraljic, M., Mandic, Z., & Duic, L. (2003). Inhibition of steel corrosion by polyaniline coatings. *Corros. Sci.*, *45*(1), 181–198.

Lacaze, P. C., Ghilane, J., Randriamahazaka, H., & Lacroix, J.-C. (2010). Electroactive conducting polymers for the protection of metals against corrosion: From micro- to nanostructured films. In: A. Eftekhari, ed., *Nanostructured Conductive Polymers.* Chichester, UK: John Wiley & Sons, Ltd.

Lacroix, J. C., Camalet, S., Aeiyach, S., Chane-Ching, K. I., Petitjean, J., Chauveau, E., & Lacaze, P. C. (2000). Aniline electropolymerization on mild steel and zinc in a two-step process. *J. Electroanal. Chem.*, *481*, 76–81.

Lee, R.-H., Lai, H.-H., Wang, J.-J., Jeng, R.-J., & Lin, J.-J. (2008). Self-doping effects on the morphology, electrochemical and conductivity properties of self-assembled polyanilines. *Thin Solid Films*, *517*, 500–505.

Lei, Y. H., Sheng, N., Hyono, A., Ueda, M., & Ohtsuka, T. (2013). Electrochemical synthesis of polypyrrole films on copper from phytic solution for corrosion protection. *Corros. Sci.*, *76*, 302–309.

Li, G., Zhang, C., Peng, H., Chen, K., & Zhang, Z. (2008). Hollow self-doped polyaniline micro/nanostructures: Microspheres, aligned pearls, and nanotubes. *Macromol. Rapid Commun.*, *29*, 1954–1958.

Liao, Y., Strong, V., Chian, W., Wang, X., Li, X.-G., & Kaner, R. B. (2012). Sulfonated polyaniline nanostructures synthesized via rapid initiated copolymerization with controllable morphology, size, and electrical properties. *Macromolecules*, *45*, 1570–1579.

Loewe, R. S., Ewbank, P. C., Liu, J., Zhai, L., & McCullough, R. D. (2001). Regioregular, head-to-tail coupled poly(3-alkylthiophenes) made easy by the GRIM method: Investigation of the reaction and the origin of regioselectivity. *Macromolecules*, *34*, 4324–4333.

Lu, H. B., Zhou, Y. Z., Vongehr, S., Hu, K., & Meng, X. K. (2011). Electropolymerization of PAN coating in nitric acid for corrosion protection of 430 SS. *Synth. Met.*, *161*(13–14), 1368–1376.

Luo, J., Jiang, S., Liu, R., Zhang, Y., & Liu, X. (2013). Synthesis of water dispersible polyaniline/poly(styrenesulfonic acid) modified graphene composite and its electrochemical properties. *Electrochim. Acta*, *96*, 103–109.

Luo, Z., Zhu, L., Zhang, H., & Tang, H. (2013). Polyaniline uniformly coated on graphene oxide sheets as supercapacitor material with improved capacitive properties. *Mater. Chem. Phys.*, *139*(2–3), 572–579.

Mahmoudian, M. R., Basirun, W. J., & Alias, Y. (2011a). Synthesis and characterization of poly(N-methylpyrrole)/TiO2 composites on steel. *Appl. Surf. Sci.*, *257*(8), 3702–3708.

Mahmoudian, M. R., Basirun, W. J., & Alias, Y. (2011b). Synthesis of polypyrrole/Ni-doped TiO2 nanocomposites (NCs) as a protective pigment in organic coating. *Prog. Org. Coat.*, *71*(1), 56–64.

Mahmoudian, M. R., Basirun, W. J., Alias, Y., & Ebadi, M. (2011). Synthesis and characterization of polypyrrole/Sn-doped TiO2 nanocomposites (NCs) as a protective pigment. *Appl. Surf. Sci.*, *257*(20), 8317–8325.

Mahmoudian, M. R., Alias, Y., & Basirun, W. J. (2012). Effect of narrow diameter polyaniline nanotubes and nanofibers in polyvinyl butyral coating on corrosion protective performance of mild steel. *Prog. Org. Coat.*, *75*(4), 301–308.

Marrion, A. R. (2004). *The Chemistry and Physics of Coatings.* Cambridge, UK: Royal Society of Chemistry.

Medrano-Vaca, M. G., Gonzalez-Rodriguez, J. G., Nicho, M. E., Casales, M., & Salinas-Bravo, V. M. (2008). Corrosion protection of carbon steel by thin films of poly (3-alkyl thiophenes) in 0.5 M H2SO4. *Electrochim. Acta*, *53*, 3500.

Michael, K., Prochaska, C., & Sazou, D. (2015). Electrodeposition of self-doped copolymers of aniline with aminobenzensulfonic acids on stainless steel. Morphological and electrochemical characterization. *J. Solid State Electrochem.*, doi: 10.1007/s10008-015-2898-4.

Mirmohsen, A., & Oladegaragoze, A. (2000). Anticorrosive properties of polyaniline coating on iron. *Synth. Met.*, *114*, 105–108.

Mondal, S. K., Prasad, K. R., & Munichandraiah, N. (2005). Analysis of electrochemical impedance of polyaniline films prepared by galvanostatic, potentiostatic and potentiodynamic methods. *Synth. Met.*, *148*(3), 275–286.

Mostafaei, A., & Zolriasatein, A. (2012). Synthesis and characterization of conducting polyaniline nanocomposites containing ZnO nanorods. *Prog. Nat. Sci. Mater. Int.*, *22*(4), 273–280.

Mostafaei, A., & Nasirpouri, F. (2013). Electrochemical study of epoxy coating containing novel conducting nanocomposite comprising polyaniline–ZnO nanorods on low carbon steel. *J. Corros. Eng. Sci. Technol.*, *48*, 513–524.

Mostafaei, A., & Nasirpouri, F. (2014). Epoxy/polyaniline–ZnO nanorods hybrid nanocomposite coatings: Synthesis, characterization and corrosion protection performance of conducting paints. *Prog. Org. Coat.*, *77*(1), 146–159.

Mrad, M., Dhouibi, L., & Triki, E. (2009). Dependence of the corrosion performance of polyaniline films applied on stainless steel on the nature of electropolymerisation solution. *Synth. Met.*, *159*(17–18), 1903–1909.

Munger, C. G., & Vincent, L. D. (1999). *Corrosion Prevention by Protective Coatings.* Houston, TX: NACE International.

Myers, R. E. (1986). Chemical oxidative polymerization as a synthetic route to electrically conducting polypyrroles. *J. Electronic Mat.*, *15*, 61–69.

Nguyen Thi Le, H., Garcia, B. R. M., Deslouis, C., & Le Xuan, Q. (2001). Corrosion protection and conducting polymers: Polypyrrole films on iron. *Electrochim. Acta*, *46*, 4259–4272.

Obaid, A. Y., El-Mossalamy, E. H., Al-Thabaiti, S. A., El-Hellag, I. S., Hermas, A. A., & Asiri, A. M. (2014). Electrodeposition and characterization of polyaniline on stainless steel surface via cyclic, convolutive voltammetry and SEM in aqueous acidic solutions. *Int. J. Electrochem. Sci.*, 9, 1003–1015.

Ocampo, C., Armelin, E., Liesa, F., Aleman, C., Ramis, X., & Iribarren, J. I. (2005). Application of a polythiophene derivative as anticorrosive additive for paints. *Prog. Org. Coat.*, *53*(3), 217–224.

Ohsaka, T., Ohnuki, Y., & Oyama, N. (1984). IR absorption spectroscopic identification of electroactive and electroinactive polyaniline films prepared by the electrochemical polymerization of aniline. *J. Electroanal. Chem.*, *161*, 399–405.

Olad, A., & Rasouli, H. (2010). Enhanced corrosion protective coating based on conducting polyaniline/zinc nanocomposite. *J. Appl. Polym. Sci.*, *115*(4), 2221–2227.

Olad, A., Barati, M., & Shirmohammadi, H. (2011). Conductivity and anticorrosion performance of polyaniline/zinc composites: Investigation of zinc particle size and distribution effect. *Prog. Org. Coat.*, *72*(4), 599–604.

Olad, A., Barati, M., & Behboudi, S. (2012). Preparation of PAN/epoxy/Zn nanocomposite using Zn nanoparticles and epoxy resin as additives and investigation of its corrosion protection behavior on iron. *Prog. Org. Coat.*, *74*(1), 221–227.

Ozyilmaz, A. T. (2006). The corrosion performance of polyaniline film modified on nickel plated copper in aqueous p-toluenesulfonic acid solution. *Surf. Coat. Technol*, *200*(12–13), 3918–3925.

Ozyilmaz, A. T., Erbil, A., & Yazici, B. (2004). Investigation of corrosion behaviour of stainless steel coated with polyaniline via electrochemical impedance spectroscopy. *Prog. Org. Coat.*, *51*(1), 47–54.

Ozyilmaz, A. T., Erbil, M., & Yazici, B. (2006a). The corrosion behaviours of polyaniline coated stainless steel in acidic solutions. *Thin Solid Films*, *496*(2), 431–437.

Ozyilmaz, A. T., Erbil, M., & Yazici, B. (2006b). The electrochemical synthesis of polyaniline on stainless steel and its corrosion performance. *Curr. Appl. Phys.*, *6*(1), 1–9.

Patil, R. C., & Radhakrishnan, S. (2006). Conducting polymer based hybrid nano-composites for enhanced corrosion protective coatings. *Prog. Org. Coat.*, *57*, 332–336.

Patil, A. O., Ikenoue, Y., Wudl, F., & Heeger, A. J. (1987). Water-soluble conducting polymers. *J. Am. Chem. Soc.*, *109*(6), 1858–1859.

Petitjean, J., Aeiyach, S., Lacroix, J. C., & Lacaze, P. C. (1999). Ultra-fast electropolymerization of pyrrole in aqueous media on oxidizable metals in a one-step process. *J. Electroanal. Chem.*, *478*, 92–100.

Petitjean, J., Tanguy, J., Lacroix, J. C., Chane-Ching, K. I., Aeiyach, S., Delamar, M., & Lacaze, P. C. (2005). Interpretation of the ultra-fast electropolymerization of pyrrole in aqueous media on zinc in a one-step process: The specific role of the salicylate salt investigated by X-ray photoelectron spectroscopy (XPS) and by electrochemical quartz crystal microbalance (EQCM). *J. Electroanal. Chem.*, *581*, 111–121.

Pournaghi-Azar, M. H., & Nahalparvari, H. (2005). Zinc hexacyanoferrate film as an effective protecting layer in two-step and one-step electropolymerization of pyrrole on zinc substrate. *Electrochim. Acta*, *50*, 2107–2115.

Prabakar, R., S. J., & Pyo, M. (2012). Corrosion protection of aluminum in LiPF6 by poly(3,4-ethylenedioxythiophene) nanosphere-coated multiwalled carbon nanotube. *Corros. Sci.*, *57*, 42–48.

Prasad, K. R., & Munichandraiah, N. (2001). Potentiodynamic deposition of polyaniline on non-platinum metals and characterization. *Synth. Met.*, *123*(3), 459–468.

Pud, A., Shapoval, G., Kamarchik, P., Ogurtsov, N., Gromovaya, V., Myronyuk, I., & Kontsur, Y. (1999). Electrochemical behavior of mild steel coated by polyaniline doped with organic sulfonic acids. *Synth. Met.*, *107*, 111–115.

Qi, K., Qiu, Y., Chen, Z., & Guo, X. (2012). Corrosion of conductive polypyrrole: Effects of environmental factors, electrochemical stimulation, and doping anions. *Corros. Sci.*, *60*, 50–58.

Qin, Q., Tao, J., & Yang, Y. (2010). Preparation and characterization of polyaniline film on stainless steel by electrochemical polymerization as a counter electrode of DSSC. *Synth. Met.*, *160*(11–12), 1167–1172.

Rammelt, U., Nguyen, P. T., & Plieth, W. (2001). Protection of mild steel by modification with thin films of polymethylthiophene. *Electrochim. Acta*, *46*(26–27), 4251–4257.

Reynolds, J., Sundaresan, N., Pomerantz, M., Basak, S., & Baker, C. (1988). Self-doped conducting copolymers: A charge and mass transport study of poly{pyrrole-CO[3-(pyrrol-1-YL) propanesulfonate]}. *J. Electroanal. Chem. Interfacial Electrochem.*, *250*, 355–371.

Riaz, U., Ahmad, S., & Ashraf, S. M. (2009). Comparison of corrosion protective performance of nanostructured polyaniline and poly(1-napthylamine)-based alkyd coatings on mild steel. *Mater. Corros.-Werkst. Korros.*, *60*, 280–286.

Rohwerder, M. (2009). Conducting polymers for corrosion protection: A review. *Int. J. Mat. Res.*, *100*(10), 1331–1342.

Ryu, H., Sheng, N., Ohtsuka, T., Fujita, S., & Kajiyama, H. (2012). Polypyrrole film on 55% Al–Zn-coated steel for corrosion prevention. *Corros. Sci.*, *56*, 67–77.

Sakhri, A., Perrin, F. X., Benaboura, A., Aragon, E., & Lamouri, S. (2011). Corrosion protection of steel by sulfo-doped polyaniline-pigmented coating. *Prog. Org. Coat.*, *72*(3), 473–479.

Santos, J. R., Mattoso, L. H. C., & Motheo, A. J. (1998). Investigation of corrosion protection of steel by polyaniline films. *Electrochim. Acta*, *43*(3–4), 309–313.

Sathiyanarayanan, S., Muthukrishnan, S., & Venkatachari, G. (2006a). Corrosion protection of steel by polyaniline blended coating. *Electrochim. Acta*, *51*, 6313–6319.

Sathiyanarayanan, S., Muthukrishnan, S., & Venkatachari, G. (2006b). Performance of poly-aniline pigmented vinyl acrylic coating on steel in aqueous solutions. *Prog. Org. Coat.*, *55*, 5–10.

Sathiyanarayanan, S., Azim, S. S., & Venkatachari, G. (2007a). A new corrosion pro-tection coating with polyaniline–TiO$_2$ composite for steel. *Electrochim. Acta*, *52*, 2068–2074.

Sathiyanarayanan, S., Azim, S. S., & Venkatachari, G. (2007b). Preparation of polyaniline–TiO2 composite and its comparative corrosion protection performance with polyani-line. *Synth. Met.*, *157*, 205–213.

Sathiyanarayanan, S., Azim, S., & Venkatachari, G. (2008). Corrosion protection of magne-sium alloy ZM21 by polyaniline-blended coatings. *J. Coat. Technol. Res.*, *5*, 471–477.

Sathiyanarayanan, S., Jeyaram, R., Muthukrishnan, S., & Venkatachari, G. (2009). Corrosion protection mechanism of polyaniline blended organic coating on steel. *J. Electrochem. Soc.*, *156*(4), C127–C134.

Sathiyanarayanan, S., Karpakam, V., Kamaraj, K., Muthukrishnan, S., & Venkatachari, G. (2010). Sulphonate doped polyaniline containing coatings for corrosion protection of iron. *Surf. Coat. Technol.*, *204*(9–10), 1426–1431.

Sazou, D. (2001). Electrodeposition of ring-substituted polyanilines on Fe surfaces from aqueous oxalic acid solutions and corrosion protection of Fe. *Synth. Met.*, *118*, 133–147.

Sazou, D., & Georgolios, C. (1997). Formation of conducting polyaniline coatings on iron surfaces by electropolymerization of aniline in aqueous solutions. *J. Electroanal. Chem.*, *429*(1–2), 81–93.

Sazou, D., Kourouzidou, M., & Pavlidou, E. (2007). Potentiodynamic and potentiostatic deposition of polyaniline on stainless steel: Electrochemical and structural studies for a potential application to corrosion control. *Electrochim. Acta*, *52*(13), 4385–4397.

Schirmeisen, M., & Beck, F. (1989). Electrocoating of iron and other metals with polypyrrole. *J. Appl. Electroctrochem*, *19*, 401–409.

Shabani-Nooshabadi, M., Ghoreishi, S. M., & Behpour, M. (2011). Direct electrosynthesis of polyaniline–montmorrilonite nanocomposite coatings on aluminum alloy 3004 and their corrosion protection performance. *Corros. Sci.*, *53*(9), 3035–3042.

Shauer, T., Joos, A., Dulog, L., & Eisenbach, C. D. (1998). Protection of iron against corrosion with polyaniline primers. *Prog. Org. Coat.*, *33*, 20–27.

Silva, J. E. P., Torresi, S. I. C., & Torresi, R. M. (2005). Polyaniline acrylic coatings for corro-sion inhibition: The role played by counter-ions. *Corros. Sci.*, *47*, 811–822.

Silva, J. E. P., Torresi, S. I. C., & Torresi, R. M. (2007). Polyaniline/poly(methylmethacrylate) blends for corrosion protection: The effect of passivating dopants on different metals. *Prog. Org. Coat.*, *58*, 33–39.

Sitaram, S. P., Stoffer, J. O., & O'Keefe, T. J. (1997). Application of conducting polymers in corrosion protection. *J. Coat. Technol.*, *69*, 65–69.

Siva, T., Kamaraj, K., & Sathiyanarayanan, S. (2014). Epoxy curing by polyaniline (PAN)—Characterization and self-healing evaluation. *Prog. Org. Coat.*, *77*, 1095–1103.

Skotheim, T. A., & Reynolds, J. R. (2007). *Conjugated Polymers: Theory, Synthesis, Properties, and Characterization* (3rd ed.). New York: CRC Press.

Souza, S. (2007). Smart coating based on polyaniline acrylic blend for corrosion protection of different metals. *Surf. Coat. Technol.*, *201*, 7574–7581.

Spinks, G. M., Dominis, A. J., Wallace, G. G., & Tallman, D. E. (2002). Electroactive conducting polymers for corrosion control—Part 2. Ferrous metals. *J. Solid State Electrochem.*, *6*(2), 85–100.

Stenger-Smith, J. D., Anderson, N., Webber, C., & Zaarei, D. (2004). Poly(2,5-bis(N-methyl-N-hexylamino) phenylene vinylene) (BAM-PPV) as replacements for chromate conver-sion coatings (CCCs). *Polym. Prepr.*, *223*, 57.

Stilwell, D. E., & Park, S. M. (1988). Electrochemistry of conductive polymers. 2. Electrochemical studies on growth properties of polyaniline. *J. Electrochem. Soc.*, *135*(9), 2254–2262.

Su, W. C., & Iroh, J. O. (2000). Electrodeposition mechanism, adhesion and corrosion performance of polypyrrole and poly(N-methylpyrrole) coatings on steel substrates. *Synth. Met.*, *114*(3), 225–234.

Tallman, D. E., Spinks, G., Dominis, A., & Wallace, G. G. (2002). Electroactive conducting polymers for corrosion control Part 1. General introduction and a review of non-ferrous metals. *J. Solid State Electrochem.*, *6*(2), 73–84.

Tan, C. K., & Blackwood, D. J. (2003). Corrosion protection by multilayered conducting polymer coatings. *Corros. Sci.*, *45*, 545–557.

Tian, Z., Yu, H., Wang, L., Saleem, M., Ren, F., Ren, P., Chen, Y., Sun, R., Sun, Y. H., & Huang, L. (2014). Recent progress in the preparation of polyaniline nanostructures and their applications in anticorrosive coatings. *RSC Adv.*, *4*, 28195–28208.

Torresi, R. M., Souza, S., Silva, J. E. P., & Torresi, S. I. C. (2005). Galvanic coupling between metal substrate and polyaniline acrylic blends: Corrosion protection mechanism. *Electrochim. Acta*, *50*, 2213–2218.

Tuken, T., Yazici, B., & Erbil, M. (2004). The electrochemical synthesis and corrosion performance of polypyrrole on brass and copper. *Prog. Org. Coat.*, *51*(2), 152–160.

Tuken, T., Yazici, B., & Erbil, A. (2005). Polypyrrole/polythiophene coating for copper protection. *Prog. Org. Coat.*, *53*, 38–45.

Ustamehmetoglu, B., Sezer, E., Kızılcan, N., Yazıcı, P., Tayyar, S., & Sezai Sarac, A. (2013). Inhibition of pyrite corrosion and photocorrosion by MEKF-R modified carbazoles. *Prog. Org. Coat.*, *76*, 533–540.

Wallace, G. G., Spinks, G. M., Kane-Maguire, L. A. P., & Teasdale, P. R. (2003). *Conductive Electroactive Polymers-Intelligent Materials Systems*. New York: CRC Press.

Waltman, R. J., & Bargon, J. (1986). Electrically conducting polymers: A review of the electropolymerization reaction, of the effects of chemical structure on polymer film properties, and of applications towards technology. *J. Can. Chem.*, *64*, 76–95.

Wei, X. L., & Epstein, A. J. (1995). Synthesis of highly sulfonated polyaniline. *Synth. Met.*, *74*(2), 123–125.

Wei, X. L., Wang, Y. Z., Long, S. M., Bobeczko, C., & Epstein, A. J. (1996). Synthesis and physical properties of highly sulfonated polyaniline. *J. Am. Chem. Soc.*, *118*(11), 2545–2555.

Weng, C.-J., Chang, C.-H., Lin, I. L., Yeh, J.-M., Wei, Y., Hsu, C.-L., & Chen, P.-H. (2012). Advanced anticorrosion coating materials prepared from fluoro–polyaniline–silica composites with synergistic effect of superhydrophobicity and redox catalytic capability. *Surf. Coat. Technol.*, *207*, 42–49.

Wessling, R. A. (1985). The polymerization of xylylene bisdialkyl sulfonium salts. *J. Polym. Sci. Polym. Symp.*, *72*, 55–66.

Wessling, B. (1994). Passivation of metals by coating with polyaniline—Corrosion potential shift and morphological changes. *Adv. Mat.*, *6*(3), 226–228.

Wicks, Z. W. J., Jones, F. N., Pappas, S. P., & Wicks, D. A. (2007). *Organic Coatings, Science and Technology*. Hoboken, NJ: John Wiley & Sons, Inc.

Xiao, R., Cho, S. I., Liu, R., & Lee, S. B. (2007). Controlled electrochemical synthesis of conductive polymer nanotube structures. *J. Am. Chem. Soc.*, *129*(14), 4483–4489.

Xing, C., Zhang, Z., Yu, L., Waterhouse, G. I. N., & Zhang, L. (2014). Anti-corrosion performance of nanostructured poly(aniline-co-metanilic acid) on carbon steel. *Prog. Org. Coat.*, *77*(2), 354–360.

Yalcinkaya, S., Tuken, T., Yazici, B., & Erbil, M. (2010). Electrochemical synthesis and corrosion behaviour of poly (pyrrole–co-o-anisidine–co-o-toluidine). *Curr. Appl. Phys.*, *10*(3), 783–789.

Yan, M. C., Vetter, C. A., & Gelling, V. J. (2010). Electrochemical investigations of polypyr-role aluminum flake coupling. *Electrochim. Acta*, *55*(20), 5576–5583.

Yang, C. F., Smyrl, W. H., & Cussler, E. L. (2004). Flake alignment in composite coatings. *J. Membrane Sci.*, *231*(1–2), 1–12.

Yang, X. G., Li, B., Wang, H. Z., & Hou, B. R. (2010). Anticorrosion performance of polyani-line nanostructures on mild steel. *Prog. Org. Coat.*, *69*(3), 267–271.

Yao, B., Wang, G., Ye, J., & Li, X. (2008). Corrosion inhibition of carbon steel by polyaniline nanofibers. *Mater. Lett.*, *62*, 1775–1778.

Yao, B., Wang, G., Li, X., & Zhang, Z. (2009). Anticorrosive properties of epoxy resin coatings cured by aniline/p-phenylenediamine copolymer. *J. Appl. Polym. Sci.*, *112*, 1988–1993.

Yue, J., Wang, Z. H., Cromack, K. R., Epstein, A. J., & MacDiarmid, A. G. (1991). Effect of sulfonic acid group on polyaniline backbone. *J. Am. Chem. Soc.*, *113*, 2665–2671.

Yue, J., Gordon, G., & Epstein, A. J. (1992). Comparison of different synthetic routes for sulphonation of polyaniline. *Polymer*, *33*, 4410–4418.

Zaarei, D., Sarabi, A. A., Sharif, F., Gudarzi, M. M., & Kassihira, S. M. (2012). A new approach to using submicron emeraldine-base polyaniline in corrosion-resistant epoxy coatings. *J. Coat. Technol. Res.*, *9*, 47–57.

Zaid, B., Aeiyach, S., Lacaze, P. C., & Takenouti, H. (1998). A two-step electropolymerization of pyrrole on Zn in aqueous media. *Electrochim. Acta*, *43*, 2331–2339.

Zarras, P., & Stenger-Smith, J. D. (2014). Electro-active polymer (EAP) coatings for corro-sion protection of metals. In: A. S. H. Makhlouf, ed., *Handbook of Smart Coatings for Materials Protection* (pp. 328–369). Cambridge, UK: Woodhead Publishing Ltd.

Zarras, P., Anderson, N., Webber, C., Irvin, D. J., Irvin, J. A., Guenthner, A., & Stenger-Smith, J. D. (2003). Progress in using conductive polymers as corrosion-inhibiting coat-ings. *Rad. Phys. Chem.*, *68*, 387–394.

Zhou, C. Q., Han, J., & Guo, R. (2009). Synthesis of polyaniline hierarchical structures in a dilute SDS/HCl solution: Nanostructure-covered rectangular tubes. *Macromolecules*, *42*, 1252–1257.

5 Characterization of Anticorrosion Properties

5.1 INTRODUCTION

Corrosion tests, interpretations, and applications of the results are considered to be a most important aspect of corrosion science and engineering. Well-planned and well-executed tests result in repeatability and reliability. These tests can be divided into laboratory tests, pilot plant tests, and field tests. The anticorrosion properties of conducting polymer coatings on active metals using laboratory tests are well documented in the literature. However, a fewer number of pilot plant and field tests were found to be conducted.

5.2 BASICS OF CORROSION RATE MEASUREMENTS

Corroding systems are not at equilibrium; therefore, thermodynamic potentials do not give an idea about the rates of corrosion. The corrosion rate can be determined as corrosion current by measuring the rate at which electrons are removed from the metal in the anodic reaction. The corrosion current can be converted to a rate of metal loss from its surface by using Faraday's law as follows (Kelly & Scully, 2003):

$$W = \frac{ItM}{nF} \tag{5.1}$$

In terms of current density, Equation 5.1 can be written as

$$W = \frac{iAtM}{nF} \tag{5.2}$$

where
   W = weight of metal in grams corroded in aqueous solution in t seconds
   I = current flow in amperes and i is the current density in amperes per cm^2
   M = atomic weight of the metal in grams/mol
   n = number of electrons/atom generated or consumed
   F = Faraday's constant = 96,500 C/mol or A.s/mol
   A = corroded area in cm^2

During corrosion, both anodic and cathodic reactions are coupled together on the electrode surface at a specific current density known as corrosion current density. However, because of the requirements of charge balance, generation rate of electrons in the anodic reaction is equal to the consumption rate of electrons in the cathodic reaction. Hence, no net current or corrosion current density can be measured.

5.2.1 Weight Loss Measurements

In this technique, samples are suspended in a solution containing the aggressive ions such as Cl-. The weight of the sample is measured at regular intervals over a long period, e.g., a year. Assuming that the change in weight represents only a loss of metal to the solution, corrosion in the simplest case, it can be converted to mol cm^{-2} s^{-1} or a corrosion current in A/cm^2 by using Faraday's equation. The corrosion rate can be calculated by using following expression (Bockris & Reddy, 2000):

$$CR\left(\frac{mm}{year}\right) = 87.6\frac{(W_1 - W_2)}{Atd}, \qquad (5.3)$$

where W_1 and W_2 are the initial and final weights before and after immersion, respectively in grams; A is the surface area in cm^2; t is total time of immersion; and d is the density in g/cm^3.

Procedures for preparing, cleaning, and evaluating corrosion test samples and conducting coupon tests in plant equipment and laboratory immersion are described in ASTM standards G1, G4, and G31. The test offers advantages such as economic method and permit analysis of corrosion products in a laboratory or on site. It requires a long immersion period to be more accurate. Despite the long duration required to conduct this test, weight loss measurements represent reality more than short time measurements do; however, such a long-term approach cannot be used to quickly determine corrosion resistance. Another limitation is the most dangerous corrosion involves internal cracking of the metal so that, with little loss to the solution, a significant loss of strength in the metal can occur. Nevertheless, few researchers used this method in their investigations. Al Dulami et al. (2011) used two inorganic pigments, TiO$_2$ and SiO$_2$, in acrylic to prepare polyaniline (PAN)-composite-based paints. The corrosion resistance of painted carbon steel samples was evaluated through visual monitoring using a digital camera after 60 days of fully immersion test in 5% NaCl as per ASTM G31 standards. At the end of the immersion period, the samples were taken out for digital photography and weight loss measurements. The weight loss measurements, as shown in Figure 5.1, revealed that the acrylic paint containing PAN–SiO$_2$ was more effective in corrosion protection compared with other pigments.

Since corrosion is an electrochemical process, it can be monitored by using electrochemical techniques. Several methods are available, as described in the following sections.

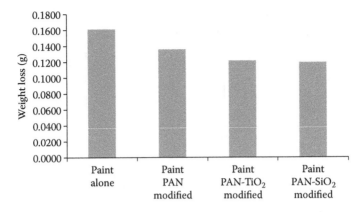

FIGURE 5.1   Weight loss measurements. (From Al Dulami, A. A. et al., *Sains Malaysiana*, 40(7), p. 762, 2011. With permission.)

5.3   DIRECT CURRENT (DC) ELECTROCHEMICAL METHODS

5.3.1   OPEN CIRCUIT POTENTIAL (HALF CELL POTENTIAL MEASUREMENTS, E_{oc})

The basic electrochemical technique in corrosion science and engineering is to measure the open circuit potential (OCP) of the corroding metal against a reference electrode and study its variation with time. It indicates the tendency of the metal or an alloy to corrode. The OCP is expressed in volts against a reference electrode such as saturated calomel electrode, which has the potential of 0.242 mV vs. the saturated hydrogen electrode (assumed potential of 0.0 V at room temperature). The more negative the potential, the higher the tendency to corrode. In general, if the OCP or corrosion potential falls to more active values, then corrosion is developing, but a shift in the positive or noble direction indicates a shifting of corrosion through passivation of the metal or the formation of insoluble corrosion products. The relative values of the OCP of the test electrode and the potentials of the expected anodic and cathodic reactions will indicate whether the corrosion reaction is under anodic, mixed, and cathodic control. Although this approach does not provide information on rate of the reaction, it serves as a useful guide to study corrosion tendency, to investigate the effects of any alteration in the test conditions, and to interpret the mechanisms. Sazou and Georgolios (1997) studied electropolymerization of aniline on iron electrodes aqueous solutions of various inorganic and organic acids under potentiostatic, galvanostatic, and potentiodynamic conditions and reported inhibitive properties of the PAN films in terms of OCP measurements. Bernard et al. (1999) defined conditions for electro deposition of PAN on iron in oxalic acid and phosphoric acids. In this investigation, it was found that the OCP of iron electrode was mentioned at values of over −250 mV/SSE and it remained in the passive state for 2000 sec in the case of a passive layer made from ferrous oxalate and for 8000 sec in the case of a passive layer obtained from phosphoric acid, as shown in Figure 5.2.

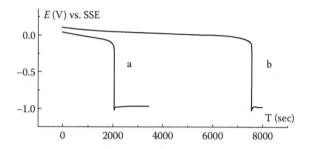

FIGURE 5.2 OCP of PAN-coated iron electrode in pH 2 solution: (a) PAN grown in 0.5 M oxalic acid + 0.1 M aniline, (b) PAN grown in 2 M phosphoric acid + 0.5 M aniline. (Reprinted from *Synth. Met.*, 102, Bernard, M., Deslouis, C., EI Moustafid, T., Hugot-Legoff, A., Joiret, S., and Tribollet, B. Combined impedance and Raman analysis in the study of corrosion protection of iron by polyaniline, 1382, Copyright 1999, with permission from Elsevier.)

Wei et al. (1995) studied base and acid doped forms of PAN coatings on cold rolled steel. Figures 5.3 and 5.4 show plots of OCP as a function of time for uncoated and coated steels in 0.1 M HCl and 3.5 wt% NaCl, respectively.

The specimens coated with 0.1 M HCl doped emeraldine base exhibited noble corrosion potential values than did the uncoated samples in acidic medium. In 3.5 wt% NaCl, however, the emeraldine base coated steel had significantly more noble corrosion potential than did the cold rolled steel coated with 0.1 M HCl doped emeraldine base and the uncoated cold rolled steel.

Sathiyanarayanan et al. (2005) investigated corrosion protection performance of conducting-polymer-based paint containing vinyl resin as a binder and PAN as the

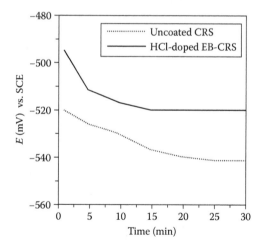

FIGURE 5.3 OCP as a function of time for uncoated and coated steels in 0.1 M HCl. CRS, cold rolled steel. (Reprinted from *Polymer*, 36, 23, Wei, Y., Wang, J., Jia, X., Yeh, J.-M., and Spellane, P., Polyaniline as corrosion protection coatings on cold rolled steel, 4536, Copyright 1995, with permission from Elsevier.)

only pigment in 3% NaCl and 0.1 N HCl solution. Figure 5.5 shows the variation in OCP with time for painted steel in 3% NaCl.

It was observed that the OCP decreased from 298 to 51 mV vs. saturated calomel electrode (SCE) in the initial stage and started to shift in noble direction after the fifth day of immersion. Afterward, the OCP value reached a steady-state value between 280

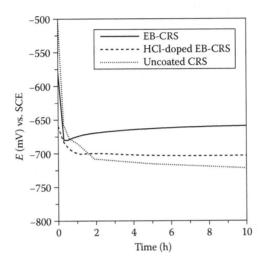

FIGURE 5.4 OCP as a function of immersion time in 3.5% NaCl. (Reprinted from *Polymer*, 36, 23, Wei, Y., Wang, J., Jia, X., Yeh, J.-M., and Spellane, P., Polyaniline as corrosion protection coatings on cold rolled steel, 4536, Copyright 1995, with permission from Elsevier.)

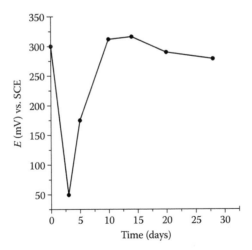

FIGURE 5.5 OCP of PAN pigmented coating on steel in 3% NaCl. (Reprinted from *Prog. Org. Coat.*, 53, Sathiyanarayanan, S., Muthukrishnan, S., Venkatachari, G., and Trivedi, D. C., Corrosion protection of steel by polyaniline pigmented paint coating, 299, Copyright 2005, with permission from Elsevier.)

and 290 mV vs. SCE. In this work, the initial decrease in OCP was assigned to the initiation of corrosion and the subsequent rise was attributed to the formation of passive film.

Syed Azim et al. (2006) prepared PAN–amino trimethylene phosphonic (PAN–ATMP) and the polymer was dispersed in epoxy resin. The variation in OCP of PAN–ATMP-coated steel in 3% NaCl with different periods of immersion is shown in Figure 5.6.

Initially, the potential shifted to active direction from −267 mV to −418 mV up to 8 days of immersion. Subsequently, the potential increased in the noble direction and reached −218 mV after 40 days of immersion. This behavior was attributed to the formation of passive layer on iron surface by PAN–ATMP.

Radhakrishnan et al. (2009) investigated the corrosion protection performance of conducting PAN–nano-TiO_2 composite paint in 3.5% NaCl. Figure 5.7 shows the variation in OCP with time for PAN–nano-TiO_2 composite painted steel.

The initial OCP of polyvinyl butyral (PVB) is found to be on cathodic side but on the positive than bare steel, whereas the coating containing conducting PAN–nano-TiO_2 composite exhibited self-healing effect. After an initial decrease in OCP, a shift in the anodic direction occurred in the second stage. This tendency increased with the increase in TiO_2 content in the composite. For PAN-4.18% TiO_2, the OCP remained on the higher anodic side of the other values even after 40 h of exposure to hot saline atmosphere. High OCP values compared with that of bare steel as well as plain PVB-coated steel clearly indicated high corrosion resistance provided by these coatings.

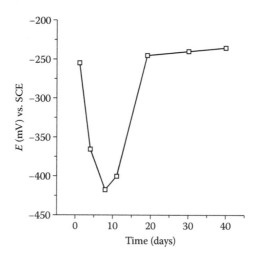

FIGURE 5.6 OCP of PAN–ATMP-pigmented coating on steel in 3% NaCl. (Reprinted from *Prog. Org. Coat.*, 56, Syed Azim, S., Sathiyanarayanan, S., and Venkatachari, G., Anticorrosive properties of PANI-ATMP polymer containing organic coatings, 157, Copyright 2006, with permission from Elsevier.)

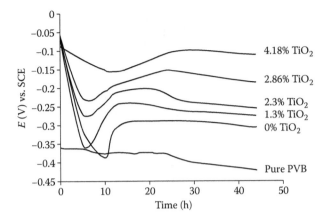

FIGURE 5.7 OCP of PAN–nano-TiO$_2$-pigmented coating on steel. (Reprinted from *Electrochim. Acta*, 54, Radhakrishnan, S., Siju, C., Mahanta, D., Patil, S., and Madras, G., Conducting polyaniline–nano-TiO$_2$ composite for smart corrosion resistant coatings, 1251, Copyright 2009, with permission from Elsevier.)

5.3.2 Polarization Techniques

5.3.2.1 Tafel Extrapolation

The corroding system is displaced from equilibrium by applying an external potential and the resultant net current is measured (potentiostatic measurements). Alternatively, the system can be displaced from equilibrium by applying known current and the resulting shit in potential can be measured (galvanostatic measurements). The difference between the applied potential and the corrosion potential, which is polarization or overpotential, is a measure of the shift in polarization potential with respect to the equilibrium potential. In anodic polarization, the anodic processes are accelerated by increasing the potential of the specimen in the positive direction. In cathodic polarization, the cathodic processes are accelerated by moving the potential in the negative direction. If the electrode is polarized from the OCP under steady-state conditions, the well-known Butler-Volmer equation can be used to find out the sum of current densities of the oxidation and reduction reactions as follows (Bockris & Reddy, 2000; Kelly & Scully, 2003):

$$i = i_0 \left\{ \exp\left(\frac{\alpha nF\eta}{RT}\right)_f - \exp\left[-(1-\alpha)\frac{nF\eta}{RT}\right]_r \right\}, \tag{5.4}$$

where $\eta = E_{\text{applied}} - E_{\text{corr}}$ is the overpotential, i_0 is the exchange current density, α is charge transfer coefficient, n is the number of electrons, R is the gas constant, T is the absolute temperature, and F is the Faraday constant.

For separate anodic and cathodic reactions, the net current density can be estimated by

$$i_a = i_0 \left(\exp \frac{\alpha n F \eta_a}{RT} \right) \qquad (5.5)$$

and

$$i_c = i_0 \left[-\exp\left(-(1-\alpha)\frac{n F \eta_c}{RT} \right) \right]. \qquad (5.6)$$

In terms of overpotential, Equations 5.5 and 5.6 can be written as follows:

$$\eta_a = \beta_a \log \frac{i_a}{i_o} \qquad (5.7)$$

and

$$\eta_c = -\beta_c \log \frac{i_c}{i_o}, \qquad (5.8)$$

where β_a and β_c, known as Tafel slopes, are given by

$$\beta_a = 2.303 \frac{RT}{\alpha n F} \qquad (5.9)$$

and

$$\beta_c = 2.303 \frac{RT}{(1-\alpha)nF}. \qquad (5.10)$$

If α is equal to 0.5 (equal proportional of electrical energy for favoring forward reaction and suppressing the reverse reaction), then the anodic and cathodic slopes are equal. An empirical relationship between the current density and over potential was given by Tafel as (Barnartt, 1978)

$$\eta = a \pm b \log i, \qquad (5.11)$$

where b is the Tafel slope of the anodic or cathodic reaction. By comparing Equations 5.7 and 5.8 with Equation 5.11, the values of constants a and b in the Tafel equation can be determined. A graph, known as Tafel plot, can be drawn showing the relationship between the overpotential and logarithmic current density, which can therefore be used to find the values of Tafel slope, corrosion potential, and corrosion current density using the extrapolation method, as described in the next paragraphs. It should be noted that the Tafel relationship is observed only when the current density is appreciable and the overvoltage is significant. A Tafel test is performed on a sample by polarizing the sample about 250 mV anodically (positive potential direction) and cathodically (negative potential direction) from the corrosion potential, E_{corr}. The potential does not have to be scanned but can be stepped in a staircase wave form. The potential scan rate can be 0.1–1.0 mV/sec. Figure 5.8 shows basis of Tafel Extrapolation method (Kelly & Scully, 2003).

As shown in Figure 5.8, the corrosion current density is obtained by extrapolation of the linear portion of the curve to E_{corr}. Assuming uniform corrosion, Faraday's law can be used to convert the corrosion density into rate of weight loss or rate of penetration. With this technique, it is possible to measure extremely low corrosion rates, and it can be used for continuous monitoring of the system under investigation. The wrong procedure, however, can lead to misleading corrosion rates. Two rules must be followed during extrapolation (Kelly et al., 2003):

1. At least one of braches of the polarization curve should exhibit Tafel behavior (i.e., linear on semi-log scale) over at least one decade.
2. The extrapolation should begin at least 50 to 100 mV away from E_{corr}.

There are certain restrictions that must be taken into account before this method can be used successfully. In many systems, Tafel region extension over a current range of at least one decade cannot be obtained because of concentration polarization and other effects. Further, the method can be applied only to systems containing one reduction process since the Tafel region gets distorted if more than one reduction

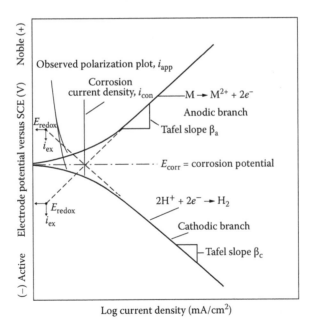

FIGURE 5.8 Tafel extrapolation. (From Kelly, R.G. and Scully, J.R., *ASM Hand Book Vol 13 A, Corrosion: Fundamentals, Testing and Protection*, Materials Park, OH: ASM International, 2003.)

process occurs (for more details on Tafel plots, please refer to ASTM G3-74 [reapproved 1981], Annual Book of ASTM Standards Vol 03-02-2001).

Wei et al. (1995) used the Tafel extrapolation method to support open circuit measurements and polarization resistance measurements of base and acid doped forms of PAN coatings on cold rolled steel. Tafel plot for uncoated cold rolled and emeraldine-base-coated cold rolled steel in 5 wt% NaCl solution is shown in Figure 5.9.

The values obtained from Figure 5.9 are recorded in Table 5.1.

From the Tafel measurements in this work, it was concluded that emeraldine-base-coated steel samples exhibited a significant reduction in corrosion rate in 5 wt% NaCl solution.

Lu et al. (1995) investigated the anticorrosion protection performance of PAN-coated steel samples exposed to artificial brine and dilute hydrochloric acid. Notations of the samples used in this work are recorded in Table 5.2.

Tafel measurements for the three sample configurations in HCl and NaCl after 8-wk exposure are shown in Figures 5.10 and 5.11.

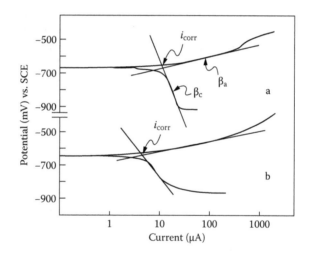

FIGURE 5.9   Tafel plot for (a) uncoated cold rolled steel and (b) emeraldine-base-coated cold rolled steel in 5 wt% NaCl solution. (Reprinted from Polymer, 36, 23, Wei, Y., Wang, J., Jia, X., Yeh, J.-M., and Spellane, P., Polyaniline as corrosion protection coatings on cold rolled steel, 4536, Copyright 1995, with permission from Elsevier.)

TABLE 5.1

Corrosion Rate Measurement Results

Sample	I_{corr} ($\mu A/cm^2$)	Corrosion Rate (mpy)
Uncoated	13	6
Emeraldine base coated	3.5	1.7

TABLE 5.2

Sample Notations

D, e/s   Control sample—epoxy-painted steel coupon. A hole 1.2 mm in diameter was drilled just
through the coating exposing clean bare surface.

D, e/dP/s   Steel sample first coated with toluene sulfonic acid doped PAN then coated with epoxy
paint. A hole 1.2 mm in diameter was drilled through the coating layers.

D, e/nP/s   Steel sample first coated with neutral PAN then coated with epoxy paint. A hole 1.2 mm in
diameter was drilled through the coating layers.

Source: Lu, W.K., Elsenbaumer, R.L., Wessling, B., *Synth. Met.*, 71, 2163–2166, 1995.

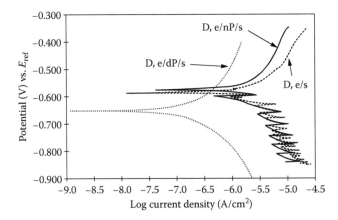

FIGURE 5.10   Tafel measurements: epoxy, neutral, and doped PAN-coated steel in HCl
solution. (Reprinted from *Synth. Met.*, 71, Lu, W.-K., Elsenbaumer, R. L., and Wessling, B.,
Corrosion protection of mild steel by coatings containing polyaniline, 2165, Copyright 1995,
with permission from Elsevier.)

The Tafel plots recorded in this work revealed that substantial reduction in corrosion current occurred in the case of doped PAN-coated sample in HCl and neutral PAN-coated sample in NaCl.

Rajagopalan and Iroh (2002) investigated the corrosion protection performance of electrochemically deposited PAN–polypyrrole (PPy) composite coatings on low-carbon steel. Figure 5.12 shows the Tafel plot of PAN formed on low-carbon steel at various applied potentials.

It was found that corrosion potential for the coated samples shifted to more positive values compared with bare steel. There was shift of almost 100 mV in the value of corrosion potential. In this work, therefore, the corrosion protection by PAN coating was assigned to prevention of cathodic reduction. Figure 5.13 shows the Tafel plot of PPy-coated low-carbon steel and control sample.

It was shown that corrosion potential for the coated samples shifted to more positive values when compared with the bare steel. There was almost a shift of 300 mV

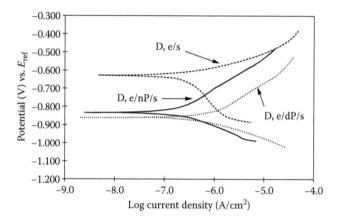

FIGURE 5.11 Tafel measurements: epoxy, neutral, and doped PAN-coated steel in NaCl solution. (Reprinted from *Synth. Met.*, 71, Lu, W.-K., Elsenbaumer, R. L., and Wessling, B., Corrosion protection of mild steel by coatings containing polyaniline, 2165, Copyright 1995, with permission from Elsevier.)

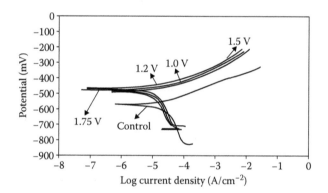

FIGURE 5.12 Tafel plots of PAN-coated low-carbon steel as a function of applied potentials. (Republished from Rajagopalan, R. and Iroh, J. O., *Surf. Eng.*, 18, p. 62, 2002. With permission of © 2002 IoM Communications. Permission conveyed through Copyright Clearance Center, Inc.)

in the value of corrosion potential. Figure 5.14 shows the Tafel plot of PAN–PPy-composite-coated low-carbon steel formed using an equimolar feed ratio of monomers and oxalic acid.

The composite coating formed at 1.5 V vs. SCE showed a shift of about 100 mV in the negative direction with respect to uncoated steel. The authors assigned this change to the possibility of change in the mechanism of corrosion protection of the composite coatings.

Samui et al. (2003) reported the results of corrosion protection of mild steel by PAN–HCl-containing paint. Corrosion protection studies were conducted under both atmospheric and underwater conditions. The painted panels were studied for

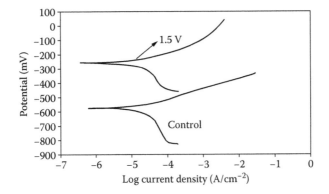

FIGURE 5.13 Tafel plots of PPy-coated low-carbon steel. (Republished from Rajagopalan, R. and Iroh, J. O., *Surf. Eng.*, 18, p. 62, 2002. With permission of © 2002 IoM Communications. Permission conveyed through Copyright Clearance Center, Inc.)

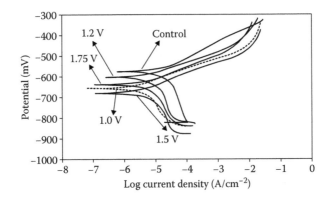

FIGURE 5.14 Tafel plots of PAN–PPy-composite-coated low-carbon steel. (Republished from Rajagopalan, R. and Iroh, J. O., *Surf. Eng.*, 18, p. 62, 2002. With permission of © 2002 IoM Communications. Permission conveyed through Copyright Clearance Center, Inc.)

their potentiodynamic behavior in NaCl solution. A Tafel plot as shown in Figure 5.15 was generated by scanning the potential from E_{corr} to ±250 mV (cathodic/anodic plots). The result of the measurement is given in Table 5.3, which revealed that higher corrosion potential values were observed for lower PAN–HCl contents.

The paints exhibited significant corrosion protection properties. Humidity cabinet, salt spray, and underwater exposure study revealed that lower PAN–HCl-containing paint protected mild steel better compared with that containing higher PAN–HCl. The paint system offered appreciable corrosion resistance without top barrier coat.

Rout et al. (2003) studied the electrochemical behavior of the coating, which was formulated by dispersing PAN powder in a medium oil alkyd resin. The corrosion rates in this work were measured by DC polarization technique in 3.5% NaCl solution using Tafel plot as shown in Figure 5.16.

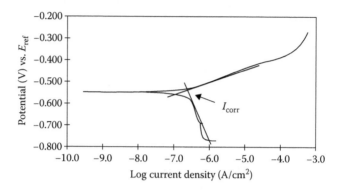

FIGURE 5.15 Tafel plot for mild steel (MS) panel coated with PAN–HCl containing paint measured in 3.5% NaCl solution. (Reprinted from *Prog. Org. Coat.*, 47, Samui, A. B., Patankar, A. S., Rangarajan, J., and Deb, P. C., Study of polyaniline containing paint for corrosion protection, 5, Copyright 2003, with permission from Elsevier.)

TABLE 5.3

Corrosion Measurements on Mild Steel Panels Coated with PAN–HCl Paint

Sample	PAN–HCl	E_{corr} (mV)	$I_{corr} \times 10^8$ (A/cm^2)	$R_p \times 10^{-6}$ (Ω cm^2)	Corrosion Rate (mpy)
PAN 0	0.0	−413.8	1.9090	0.6010	0.009
PAN 1	0.1	−283.6	0.0224	139.6	0.000
PAN 2	0.5	−347.6	1.9500	0.0874	0.018
PAN 3	1.0	−351.8	0.6377	2.6230	0.008
PAN 4	3.0	−382.0	0.4967	7.3980	0.006
PAN 5	5.0	−364.6	1.3820	0.9774	0.017
PAN 6	10.0	−347.9	1.2780	0.7202	0.016
PAN 7	20.0	−474.3	25.23	2.030	0.314
Coal tar epoxy	0.0	−524.7	19.27	0.4164	0.088

Source: Reprinted from *Prog. Org. Coat.*, 47, Samui, A. B., Patankar, A. S., Rangarajan, J., and Deb, P. C., Study of polyaniline containing paint for corrosion protection, 5, Copyright 2003, with permission from Elsevier.

It was found from the Tafel plot in this work that the corrosion rate for the PAN-coated steel was significantly lower, i.e., 0.132, mpy compared with that of bare steel, i.e., 3.606 mpy.

5.3.2.2 Potentiodynamic Measurements

The potentiodynamic method is used to study the passivation tendency of a metal or an alloy in a given solution. It is possible to determine whether the passivation is spontaneous or needs polarization to induce passivation. The degree of passivation and the stability of passive film can be assessed by measuring the passive region

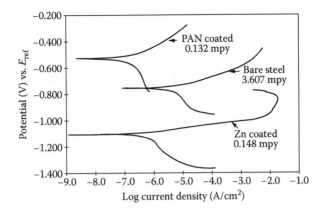

FIGURE 5.16 Tafel plot for PAN-coated, bare, and zinc-coated steel in 3.5 wt% NaCl solution. (Reprinted from *Surf. Coat. Technol.*, 167, Rout, T. K., Jha, G., Singh, A. K., Bandyopadhyay, N., and Mohanty, O. N., Development of conducting polyaniline coating: A novel approach to superior corrosion resistance, p. 20, Copyright 2003, with permission from Elsevier.)

current and the transpassivation potential. The potentiodynamic anodic scan uses a potential scan beginning at −1.5 V vs. OCP and scanning at 5 mV/sec in a positive direction. The final potential is generally set at 4 V vs. OCP. More details can be found in ASTM standard G3 (Kelly & Scully, 2003).

Potentiodynamic measurements for the three PAN-coated steel samples (notations as per Table 5.2) in HCl and NaCl are shown in Figures 5.17 and 5.18.

The potentiodynamic scans shown in Figures 5.17 and 5.18 indicated strong passivation tendency for the D,e/dP/s and D,e/nP/s samples in HCl, while there was no

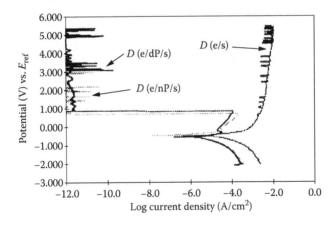

FIGURE 5.17 Potentiodynamic measurements: epoxy, neutral, and doped PAN-coated steel in HCl solution. (Reprinted from *Synth. Met.*, 71, Lu, W.-K., Elsenbaumer, R. L., and Wessling, B., Corrosion protection of mild steel by coatings containing polyaniline, 2165, Copyright 1995, with permission from Elsevier.)

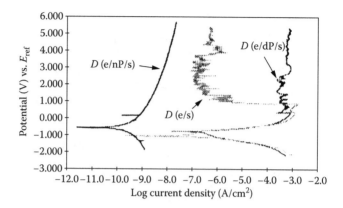

FIGURE 5.18 Potentiodynamic measurements: epoxy, neutral, and doped PAN-coated steel in NaCl solution. (Reprinted from *Synth. Met.*, 71, Lu, W.-K., Elsenbaumer, R. L., and Wessling, B., Corrosion protection of mild steel by coatings containing polyaniline, 2166, Copyright 1995, with permission from Elsevier.)

tendency for passivation with the control sample. The samples coated with PAN did not exhibit passivation in NaCl. Nevertheless, the sample coated with neutral PAN did exhibit a significant reduction in corrosion current density, as well as pronounced cathodic and anodic polarization, compared with the sample coated with doped PAN and control sample These results provided an evidence of the significant corrosion protection by galvamnically coupled PAN to exposed steel and that the level of corrosion protection was affected by the form of PAN and the nature of corrosive environment. Similar potentiodynamic measurements carried out on bare, PAN-coated, and zinc-coated samples are shown in Figure 5.19 (Rout et al., 2003).

It was concluded in this work that the catalytic behavior of the PAN coating was responsible for the formation of the passive oxide layer at shorter duration (Rout et al., 2003).

Chaudhari et al. (2010) synthesized poly(*o*-toluidine) (POT)/CdO nanoparticle composite film on mild steel from aqueous tartrate solution using cyclic voltammetry and evaluated its corrosion protection performance using potentiodynamic polarization. The potentiodynamic measurements on POT-coated, bare, and POT/CdO-coated steel in 3 wt% NaCl solution are shown in Figure 5.20.

These studies revealed that the incorporation of CdO nanoparticles in the POT matrix improved the corrosion protection properties. The protection efficiencies (PES) of POT and POT/CdO nanoparticle composite coatings were found to be 81% and 97%, respectively (Chaudhari et al., 2010).

5.3.2.3 Cyclic Polarization

Cyclic polarization is used to study the pitting tendency of metals and alloys such as stainless steel. Potential scan is started in an anodic direction from a potential near the corrosion potential. When the measured current exhibits a large increase or reaches a specified value, the scan direction is reversed to the cathodic direction. The potential at which the measured current increases significantly on the forward sweep is

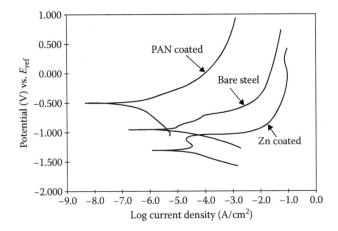

FIGURE 5.19 Potentiodynamic measurements: PAN-coated, bare, and zinc-coated steel in 3.5 wt% NaCl solution. (Reprinted from *Surf. Coat. Technol.*, 167, Rout, T. K., Jha, G., Singh, A. K., Bandyopadhyay, N., and Mohanty, O. N., Development of conducting polyaniline coating: A novel approach to superior corrosion resistance, 21, Copyright 2003, with permission from Elsevier.)

considered to be pitting potential, E_{pit}. The potential where the loop closes on the reverse scan is the protection potential, E_{prot}. New pits initiate at a potential above the pitting potential. Between the pitting potential and the protection potential, existing pits can grow. The hysteresis loop between the forward and backward scans is an indication of pitting formation. The larger the loop, the greater is the pitting tendency. If

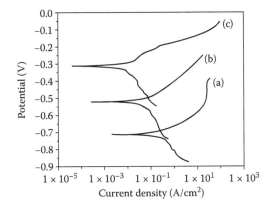

FIGURE 5.20 Potentiodynamic measurements: (a) uncoated mild steel, (b) POT-coated mild steel, and (c) POT-CdO nano particle composite coated mild steel in 3 wt% NaCl solution. (With kind permission from Springer Science+Business Media: *J. Coat. Technol. Res.*, Synthesis and corrosion protection aspects of poly(o-toluidine)/CdO nanoparticle composite coatings on mild steel, 7(1), 2010, 125, Chaudhari, S., Gaikwad, A. B., Patil, P. P., © FSCT and OCCA 2009.)

the loops do not close, E_{prot} can be determined by extrapolating the reverse scan to an infinitesimal current. When the loops are close together, the pitting tendency is small. If E_{prot} is greater than E_{pit}, there may be no tendency to pit. More details can be found in ASTM G61 (Kelly & Scully, 2003). The cyclic potentiodynamic polarization curves recorded in aqueous 3% NaCl solution for uncoated 304-stainless steel and poly(o-ethylaniline)-coated steel are shown in Figure 5.21 (Chaudhari et al., 2007).

In the case of bare sample, approximately the same values of the corrosion potential (E_{corr} = −0.214 V) and the protection potential (E_{prot} = −0.216 V) indicated that the pitting occurred at the OCP in the aqueous 3% NaCl. The absence of hysteresis loop in case of poly(o-ethylaniline)-coated steel revealed its ability to resist localized corrosion (Chaudhari et al., 2007).

5.3.2.4 Linear Polarization Resistance

When polarizing from the corrosion potential with respect to anodic or cathodic current density, the overpotential expressions given by Equations 5.7 and 5.8 become

$$\eta_a = \beta_a \log \frac{i_a}{i_{corr}} \text{ and} \tag{5.12}$$

$$\eta_c = -\beta_c \log \frac{i_c}{i_{corr}}. \tag{5.13}$$

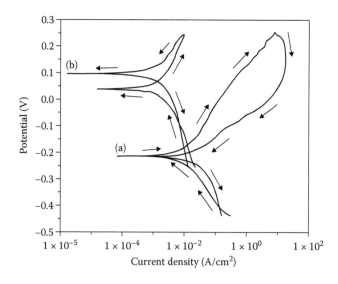

FIGURE 5.21 The cyclic potentio dynamic polarization curves recorded in aqueous 3% NaCl solution for (a) uncoated 304-stainless steel and (b) poly(o-ethylaniline)-coated steel. (Reprinted from *Prog. Org. Coat.*, 58, Chaudhari, S., Sainkar, S. R., and Patil, P. P., Poly(o-ethylaniline) coatings for stainless steel protection, p. 60, Copyright 2007, with permission from Elsevier.)

Recall that $\eta_a = E - E_{corr} > 0$ and $\eta_c = E - E_{corr} < 0$ represent the potential changes from the corrosion potential. Solving the equations for the anodic and cathodic current densities gives, respectively,

$$i_a = i_{corr} \exp\left(2.303\frac{E - E_{corr}}{\beta_a} \right) \text{ and} \qquad (5.14)$$

$$i_c = i_{corr} \exp\left(-2.303\frac{E - E_{corr}}{\beta_c} \right). \qquad (5.15)$$

Assuming that the applied current density is $i = i_a - i_c$ and substituting Equations 5.14 and 5.15 into this expression give the Butler-Volmer equation that quantifies the kinetics of the electrochemical corrosion reactions.

$$i_a = i_{corr} \left[\exp\left(2.303\frac{E - E_{corr}}{\beta_a} \right) - \exp\left(-2.303\frac{E - E_{corr}}{\beta_c} \right) \right] \qquad (5.16)$$

Deriving 5.16 equation with respect to the applied potential,

$$\frac{di}{dE} = 2.303 i_{corr} \left[\beta_a^{-1} \exp\left(2.303\frac{E - E_{corr}}{\beta_a} \right) - \beta_c^{-1} \exp\left(2.303\frac{E - E_{corr}}{\beta_c} \right) \right] \qquad (5.17)$$

Evaluating Equation 5.17 at $E = E_{corr}$ results in the following expression:

$$\frac{di}{dE} = 2.303\frac{i_{corr}(\beta_a + \beta_c)}{\beta_a\beta_c}. \qquad (5.18)$$

Hence, the polarization resistance can be given by

$$R_p = \frac{dE}{di} = \frac{\beta_a\beta_c}{2.303\, i_{corr}\, (\beta_a + \beta_c)} \qquad (5.19)$$

$$R_p = \frac{\beta}{i_{corr}} \qquad (5.20)$$

Equation 5.20, known as the Stern Geary equation (Stern & Geary, 1957), was originally formulated for the region in the vicinity of the corrosion potential where a linear dependence of potential on applied current existed for a corroding metal. It reveals that the polarization resistance is inversely proportional to corrosion current density. Taking the logarithm of this equation, it is seen that $\log i_{corr}$ vs. $\log R_p$ is linear with a slope of -1 and has the intercept $\log B$ as shown in Figure 5.22.

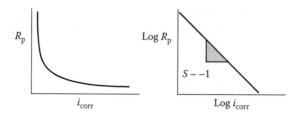

FIGURE 5.22   Nonlinear and linear relationships between polarization resistance and corrosion current density.

$$\log R_p = \log B - \log i_{corr} \tag{5.21}$$

It reveals that $\log i_{corr}$ vs. $\log R_p$ is linear with a slope of -1 and has the intercept $\log B$, and therefore, the equation forms the basis of linear polarization resistance method.

A test is performed on a sample by polarizing the sample about 10–20 mV from the corrosion potential, E_{corr}. A slow scan rate such as ~0.125 mV/sec is used to reach the final potential. A graph for $E_{app} - E_{corr}$ is plotted as a function of corrosion current, as shown in Figure 5.23. The technique permits accurate measurements of extremely low corrosion rates such as less than 0.1 mpy. The corrosion rate can be calculated using the following equation (Kelly & Scully, 2003):

$$C_{Rate} = 0.129 \; i_{corr} \; \frac{\text{Equivalent Weight}}{Ad}. \tag{5.22}$$

Corrosion rate has a unit of mpy (milli-inch per year) if corroded area (A) is measured in cm^2, density (d) in g cm^{-3}, equivalent weight in gram per equivalent and corrosion current density i_{corr} in μA/cm^2.

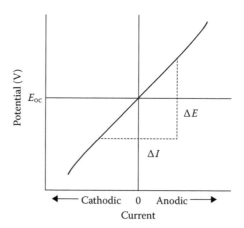

FIGURE 5.23   Linear polarization resistance method.

5.4   ALTERNATING CURRENT (AC) ELECTROCHEMICAL METHOD: ELECTROCHEMICAL IMPEDANCE SPECTROSCOPY

Electrochemical impedance spectroscopy (EIS) has been used extensively by material scientists, particularly in the last two decades, because of its ability to differentiate various electrochemical processes occurring at a corroding metallic interface. The technique has proved to be of particular advantage in conducting polymer coating research, where researchers have been able to model the corrosion reactions occurring under intact as well as damaged coating and to elucidate on the probable mechanisms of corrosion protection (Tallman et al., 2002).

5.4.1   THEORY

Electrical resistance is the ability of a circuit to oppose the passage of direct current, mathematically expressed as

$$R = \frac{V_{DC}}{I_{DC}}, \tag{5.23}$$

where V_{DC} is the applied potential in volts, I_{DC} is the current flowing through the circuit in amperes, and R is the resistance in ohms. Impedance is the ability of a circuit to oppose the passage of alternating current, mathematically expressed as

$$Z = \frac{V_{AC}}{I_{AC}}, \tag{5.24}$$

where V_{AC} is the applied potential in volts (rms), I_{AC} is the current flowing through the circuit in amperes (rms), and Z is the impedance in ohms. An AC potential is a potential with a sinusoidal wave form applied to it. It can be described by the following equation (Colreavy & Scantlebury, 1995; Kelly & Scully, 2003):

$$V_t = V_0 \sin \omega t, \tag{5.25}$$

where V_t is the applied potential at time t, V_0 is the amplitude of the applied potential signal, ω ($2\pi f$) is the radial frequency, and t is time. The amplitude of this signal must be low, as the simple linear relationship relating resistance to current and voltage, given by ohm's law, becomes nonlinear with more complex circuits. When an AC potential is applied to a circuit, the resulting current remains sinusoidal. However, it may be in phase or out of phase with the applied potential signal by an amount, the phase angle Φ, which depends on the circuit parameters. It can be given by the following expression:

$$I_t = I_0 \sin(\omega t + \phi), \tag{5.26}$$

where I_t is the resulting current, I_0 is the amplitude of the resulting current, ω ($2\pi f$) is the radial frequency, and t is time. Using Ohm's law, impedance of the system can be given by the following equation:

$$Z(\omega) = \frac{V_0 \sin \omega t}{I_0 \sin(\omega t + \phi)} = Z_0 \frac{\sin \omega t}{\sin(\omega t + \phi)} . \qquad (5.27)$$

The impedance is therefore a vector quantity containing both a magnitude and a direction, i.e., a phase shift between voltage and current. It is possible to express resulting current and applied potential as complex functions:

$$I_t = I_0 \sin(j\omega t + j\phi) \text{ and} \qquad (5.28)$$

$$V_t = V_0 \sin j\omega t. \qquad (5.29)$$

Taking the ratio of these equations, the impedance can be given as

$$Z(\omega) = \frac{V_0 \sin j\omega t}{I_0 \sin(j\omega t - j\phi)} = |z| \exp(j\phi). \qquad (5.30)$$

Using Euler's relationship, $\exp(j\phi) = \cos \phi + j \sin \phi$, the equation becomes

$$= |z|(\cos \phi + j \sin \phi) \qquad (5.31)$$

$$= Z' + jZ'' \qquad (5.32)$$

where Z' and Z'' are real and imaginary components of the impedance, respectively, and $j = \sqrt{-1}$. In the complex plane diagram, the impedance of a single frequency can be represented by a vector of length with argument Φ as shown in Figure 5.24.

By using analytical geometry, the modulus $|z|$ and phase angle Φ can be given by the following equations:

$$R_e(Z) = |z| \cos \phi, \qquad (5.33)$$

$$I_m(z) = |z| \sin \phi, \qquad (5.34)$$

$$|z| = \sqrt{(Z'^2 + Z''^2)} \text{ and } \phi = \tan(Z'/Z''). \qquad (5.35)$$

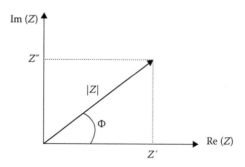

FIGURE 5.24   Real and imaginary components of impedance vector.

If a sinusoidal potential is applied across a resistance, R, then the resulting current is in phase with the applied potential for all frequencies. In this case, the magnitude of the impedance is $Z = R$ and the phase $\Phi = 0$ for all frequencies. This is shown in Figure 5.25 on a plot of real and imaginary components as a point on the real axis.

If a sinusoidal potential is applied across a pure capacitor, then the phase shift occurs as shown in Figure 5.26.

For a capacitor, $V = Q/C$, and since $I = dQ/dT$, impedance in this case can be given by following expression:

$$Z(\omega) = \frac{V_0}{(\omega C)V_0} \tag{5.36}$$

$$= \frac{1}{(\omega C)}. \tag{5.37}$$

Thus, the impedance is now dependent on frequency. As the frequency increases, the magnitude of impedance decreases, as shown in Figure 5.27.

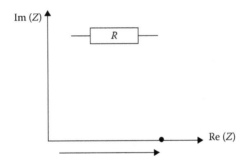

FIGURE 5.25 Impedance spectrum for resistance.

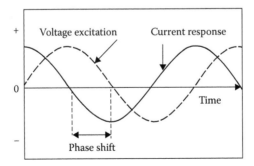

FIGURE 5.26 Relationship between voltage across a capacitor and current through the capacitor.

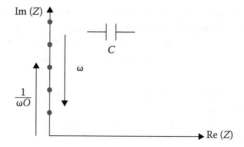

FIGURE 5.27   Impedance spectrum for capacitor.

The effect of combining both resistance and capacitance in series or parallel can be deduced as shown in Figures 5.28 and 5.29, respectively.

Thus, an electrochemical interface can be imagined as behaving as a circuit consisting of passive electrical components such as resistors and capacitors. Impedance data, therefore, can be analyzed by modeling the electrochemical processes and then using electrical equivalent circuits.

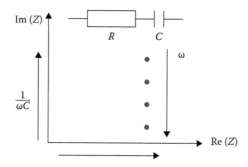

FIGURE 5.28   Series resistance and capacitor.

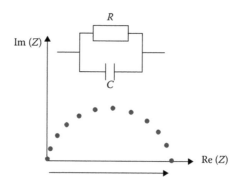

FIGURE 5.29   Parallel resistance and capacitor.

5.4.2 EXPERIMENT AND DATA ANALYSIS

The impedance measurements are performed over large frequency ranges, typically from 100 KHz to 0.001 Hz, using amplitude signal voltage range of 5 mV to 50 mV rms. For electro coating systems, a classic value of 20 mV rms is chosen to characterize intact coatings. However, for impedance measurements on scratched samples, the response of the exposed metal being dominant, a lower value of the perturbation, such as 5 mV rms, can be selected. The obtained impedance data can be represented by two types of diagrams: Nyquist plot and Bode plot. An ideal Nyquist plot showing real and imaginary components of impedance measured at different frequencies is shown in Figure 5.30.

In electrochemistry, the imaginary impedance is always capacitive and therefore negative, as shown in the Figure 5.34. The plot is a semicircle with the polarization resistance R_p as its diameter. However, the plot deviates from the semicircular shape in practice. The high-frequency intercept on real axis is the solution resistance R_s, whereas the low-frequency intercept (approaching DC signal) on the real axis is the total impedance $R_{ct} + R_s$. Therefore, subtracting the high-frequency intercept from the low-frequency intercept gives value of the charge transfer resistance R_{ct} or the polarization resistance R_p, which then can be used in corrosion rate calculations. A typical Bode plot showing the variation of impedance $-\log |z|$ and phase angle Φ as a function of the logarithm of the frequency is shown in Figure 5.31.

The solution resistance and charge transfer resistance can be determined from the values of $|z|$ for $\omega \to 0$ and $\omega \to$ infinity. At intermediate frequencies, the plot of log $|z|$ should be a straight line with slope -1. The double-layer capacitance can therefore be estimated from extrapolation of the absolute impedance curve at intermediate frequencies to the impedance axis, as shown in Figure 5.35. At the high and low frequency ends, where the behavior is like a resistor, the phase angle is 0. It increases as the imaginary component increases and becomes maximum for medium frequency. Both the plots should give the same solution and polarization resistance values. The Nyquist and Bode plots shown in Figures 5.30 and 5.31, respectively, can be modeled by electrical circuit as shown in Figure 5.32.

The circuit shown in Figure 5.32 is known as Randles circuit. R_s is the uncompensated resistance or solution resistance between the working electrode and the reference electrode. The double-layer capacitance (C_{dl}) represents charge stored in the

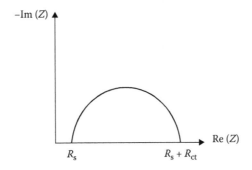

FIGURE 5.30 Nyquist plot.

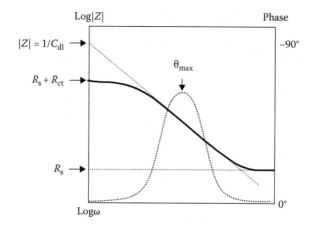

FIGURE 5.31   Bode plots showing the variation of impedance (log Z) and phase angle (Φ) with respect to changes in frequency.

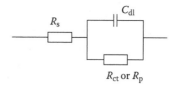

FIGURE 5.32   Equivalent electrical circuit for an electrochemical interface.

double layer at the working electrode–electrolyte interface. The resistance to charge transfer across the double-layer capacitor is given by charge transfer resistance R_{ct} or the polarization resistance R_p. It can be related to the rate of anodic and cathodic reactions. With reference to polarization of electrode in an electrolyte, double-layer capacitance C_{dl} and charge transfer resistance R_{ct} represent activation polarization and concentration polarization, respectively. Higher values of charge transfer resistance indicate more corrosion resistance offered by the metal in the electrolyte under investigation. For a carbon steel in sea water, charge transfer resistance is $10^4 \; \Omega \; cm^2$. An intact organic coating on metal behaves electrically as a dielectric and can be represented by a parallel plate capacitor. Initially, the coating does not have conducting paths or pores. The value of coating capacitance C_c in this case can be determined by using the following equation:

$$C_c = \varepsilon_0 \varepsilon_r \frac{A}{t},$$
(5.38)

where ε_0 is the permittivity of free space and ε_r is the relative permittivity or dielectric constant, A is the interface surface area, and t is the coating thickness.

The coating capacitance changes as the electrolyte penetrates through the coating and reaches the metallic surface. The volume percentage of absorbed water can be estimated by using the following relation (Brasher & Kingbury, 1954):

$$V = \frac{\log(C_t/C_0)}{\log \varepsilon}, \tag{5.39}$$

where C_t is coating capacitance at the time of measurement, C_0 is the coating capacitance at $t = 0$, and ε is the dielectric constant of water. A Nyquist plot for damaged polymer coating on the metal and equivalent electrical circuit is shown in Figure 5.33.

As shown in Figure 5.33, the circuit now involves two parallel capacitors and resistors in series along with the solution resistance. Pore resistance, which relates to diffusion of electrolyte through the coating via the pore, should be taken into account. The pore resistance is the ionic resistance of coating inversely proportional to the area of pores or surface defects. In general, the pore resistance of an organic coating is $10^{10}\ \Omega\ \mathrm{cm}^2$ and corrosion occurs if it decreases below $10^6\ \Omega\ \mathrm{cm}^2$. Thus, it is an indicative parameter for the degradation of the coating. However, the pore resistance can increase with time because of blockage of the pores by corrosion products. When the rate of diffusion of species through pores becomes lower than the charge transfer processes, a tail of 45° appears in the low-frequency region of the Nyquist plot, as shown in Figure 5.34. The tail bends toward the real axis for lower frequencies. An

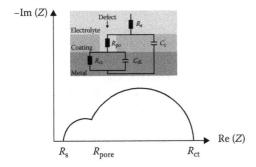

FIGURE 5.33 Nyquist plot for damaged polymer coating on the metal and equivalent electrical circuit.

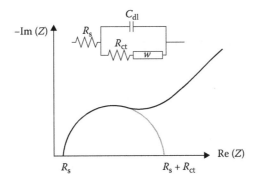

FIGURE 5.34 Nyquist plot containing Warburg impedance and equivalent electrical circuit.

additional resistance known as Warburg's impedance is added in the circuit to represent this process or when the corrosion products control the corrosion rate. Warburg impedance appears as a diagonal line with a slope of 0.5 on a Bode plot.

Ideal capacitors may not be able to represent experimental data in certain cases, such as nonuniform coatings, in homogeneous distribution of current, etc. In such cases, constant phase element (CPE) is used, as shown in Figure 5.35.

CPE can be given by the following equation

$$CPE = \frac{1}{z(2nf)^\theta},$$
(5.40)

where $0 < \theta < 1$ depends upon phase angle. For an ideal capacitor, the phase angle in bode plot is $-90°$ and $\theta = 1$.

Advantages of the Nyquist plot include simple representation of the charge transfer resistance as diameter of the semicircle and the shape does not change when the ohmic resistance is changed. However, Nyquist plots do not reveal frequency information, whereas Bode plots give a description of the frequency dependence of the electrochemical parameters. Although the technique provides valuable information, it is not without limitations—limitations on samples, in the frequency range, instrumentation used, and data analysis capabilities. Also, interpretations of the results becomes difficult when the equivalent circuit is not known or a number of circuits are possible for a given spectrum. More details can be found in ASTM G106 and ASTM 2005b.

Li et al. (1997) studied the ability of PAN (proprietary dispersion supplied by Zipperling, Germany) to act as a protective coating for mild steel corrosion in saline and hydrochloric acid by EIS. Nyquist plots of mild steel coated with PAN, insulating top coat alone, and PAN with insulating top coat in 1 M HCl for 6 h are shown in Figure 5.36.

The EIS results indicated that the PAN primer was a poor barrier and that rapid proton diffusion occurred, which was responsible for dissolving the passive layer at a rate faster than the repassivation effect. The impedance of the top coat was a near perfect semicircle, a characteristic of an impervious insulating coating. The impedance plot of PAN with the top coat showed the classical pattern of a charge transfer reaction rate limited by the diffusion of species at low frequencies. The appearance of a straight line, i.e., Warburg impedance, confirmed the occurrence of redox reactions within the composite film. It revealed that the top coat opposed the diffusion of corrosive species to the PAN–metal interface and reduced corrosion. In addition, this

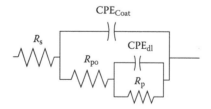

FIGURE 5.35   CPE in an equivalent electrical circuit.

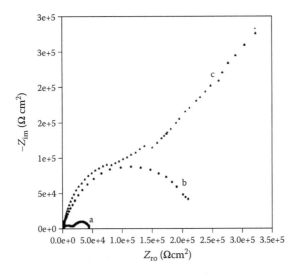

FIGURE 5.36 Nyquist plots of mild steel coated with (a) PAN, (b) insulating top coat alone, and (c) PAN with insulating top coat in 1 M HCl for 6 h. (Reprinted from *Synth. Met.*, 88, Li, P., Tan, T. C., and Lee, J. Y., Corrosion protection of mild steel by electroactive polyaniline coatings, p. 241, Copyright 1997, with permission from Elsevier.)

layer precluded passive layer dissolution and helped to maintain the passive state. A suggested model in this work for noninteracting layers of electroactive polymer and top coat is shown in Figure 5.37.

Kinlen et al. (1997) used PAN emeraldine salt powder dispersed in a polymer binder as a primer and barrier top coat. The circuit used in this research is shown in Figure 5.38. It included resistance due to oxide formation in the defect and PAN

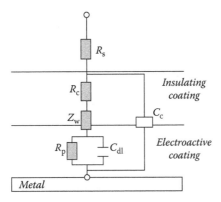

FIGURE 5.37 Circuit for interacting layers of electroactive polymer and top coat. (Reprinted from *Synth. Met.*, 88, Li, P., Tan, T. C., and Lee, J. Y., Corrosion protection of mild steel by electroactive polyaniline coatings, p. 241, Copyright 1997, with permission from Elsevier.)

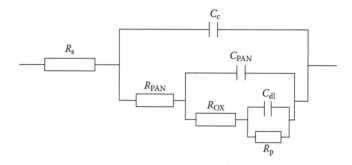

FIGURE 5.38 Equivalent circuit used to model impedance data for PAN emeraldine salt dispersed in polymer binder and epoxy top coat in 3% NaCl. (Reprinted from *Synth. Met.*, 85, Kinlen, P. J., Silverman, D., and Jeffreys, C., Corrosion protection using polyaniline coating formulation, p. 1327, Copyright 1997, with permission from Elsevier.)

capacitance. These researchers used double-layer capacitance and polarization resistance in parallel instead of Warburg impedance.

Bernard et al. (1999) used combined impedance and Raman analysis in the study of corrosion protection of iron by PAN. Figure 5.39 shows an impedance diagram for a PAN-coated iron electrode in 0.1 M NaCl between 0 and 7 days of immersion.

The impedance data showed, in addition to the high frequency loop, a long diffusional tail appearing as a straight line with a slope around 22.5⁰ with respect to the real axis and over several frequency decades. The impedance results were consistent with the Raman spectroscopy data. Both techniques indicated that the film displayed electronic conductivity at higher pH values which was unexpected.

Sathiyanarayanan et al. (2005) studied the corrosion protection performance of PAN pigmented coating on steel in 3% NaCl and 0.1 N HCl by using EIS. The Nyquist plots of PAN pigment coated steel at different immersion times in 3% NaCl are shown in Figure 5.40.

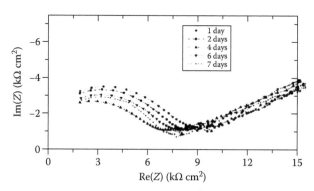

FIGURE 5.39 Impedance spectrum for a PAN-coated iron electrode in 0.1 M NaCl between 0 and 7 days of immersion. (Reprinted from *Synth. Met.*, 102, Bernard, M., Deslouis, C., El Moustafid, T., Hugot-Legoff, A., Joiret, S., and Tribollet, B., Combined impedance and Raman analysis in the study of corrosion protection of iron by polyaniline, 1382, Copyright 1999, with permission from Elsevier.)

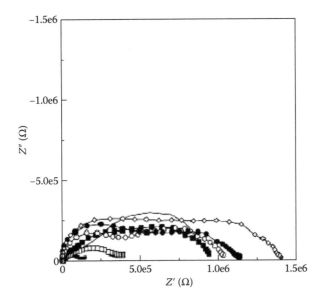

FIGURE 5.40 Impedance behavior of PAN pigmented paint coating on steel in 3% NaCl. (Reprinted from *Prog. Org. Coat.*, 53, Sathiyanarayanan, S., Muthukrishnan, S., Venkatachari, G., and Trivedi, D. C., Corrosion protection of steel by polyaniline pigmented paint coating, p. 299, Copyright 2005, with permission from Elsevier.)

The impedance data shown in Figure 5.40 were analyzed by using the same circuit shown in Figure 5.33 except that the coating resistance parameter was used instead of pore resistance. The initial decrease and subsequent increase in the coating resistance and coating capacitance were attributed to the protective properties of the PAN coating. The initial decrease in the charge transfer resistance and its subsequent increase was assigned to the dissolution of steel and the formation of passive film on the surface. The Nyquist plots of PAN-pigment-coated steel at different immersion times in 0.1 N HCl are shown in Figure 5.41.

Semicircle, which was observed in this work, revealed that the coating exhibited capacitive tendency, and this tendency increased with immersion time. Impedance data shown in Figure 5.41 was analyzed using the circuit shown in Figure 5.42.

Tuken et al. (2006) electrochemically synthesized PPy coating on mild steel from phenylphosphonic acid solution and used impedance spectroscopy to investigate corrosion protection performance. A Nyquist plot obtained in this work for bare mild steel electrode in 3.5% NaCl is shown in Figure 5.43.

Since there was no coating on the electrode and a passivation could not be expected, the diameter gave the value of the charge transfer resistance against corrosion. The R_{ct} value thus obtained was put in the Stern Geary equation for obtaining the corrosion current. Figure 5.44 shows the impedance results of coated steel samples as a function of immersion time.

After 8 h of exposure time, the Nyquist plot was observed to be made up of two different semicircles. The semicircle at high frequency was magnified and

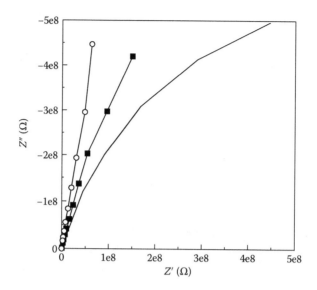

FIGURE 5.41   Impedance behavior of PAN pigmented paint coating on steel in 0.1 N HCl. (Reprinted from *Prog. Org. Coat.*, 53, Sathiyanarayanan, S., Muthukrishnan, S., Venkatachari, G., and Trivedi, D. C., Corrosion protection of steel by polyaniline pigmented paint coating, p. 300, Copyright 2005, with permission from Elsevier.)

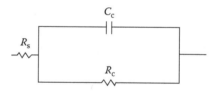

FIGURE 5.42   Equivalent circuit used to analyze the impedance data in Figure 5.45. (Reprinted from *Prog. Org. Coat.*, 53, Sathiyanarayanan, S., Muthukrishnan, S., Venkatachari, G., and Trivedi, D. C., Corrosion protection of steel by polyaniline pigmented paint coating, p. 299, Copyright 2005, with permission from Elsevier.)

given in the square window within the Nyquist diagram. This region contained the information about the charge transfer process occurring at the bottom of the pores of polymer coating. The second semicircle or, more precisely, the depressed semicircle, included information about the polymer film and the passive oxide layer formed at the interface. The charge transfer resistance and the film resistance values were obtained from Bode plot and circuit shown in Figure 5.45 and recorded in Table 5.4.

It was observed from the Nyquist and bode plots that the charge transfer resistance values increased with time and the corresponding semicircle became partially seen in Nyquist plots. In the case of nonconducting barrier coating, a depressed semicircle of decreasing diameter with increasing exposure time is observed.

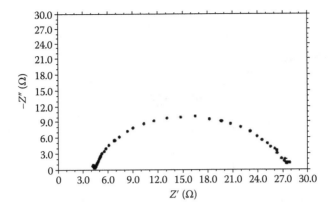

FIGURE 5.43 The Nyquist plot obtained for mild steel electrodes in 3.5% NaCl. (Reprinted from *App. Sur. Sci.*, 252, Tuken, T., Yazici, B., and Erbil, M., The corrosion behavior of polypyrrole coating synthesized in phenylphosphonic acid solution, p. 2314, Copyright 2006, with permission from Elsevier.)

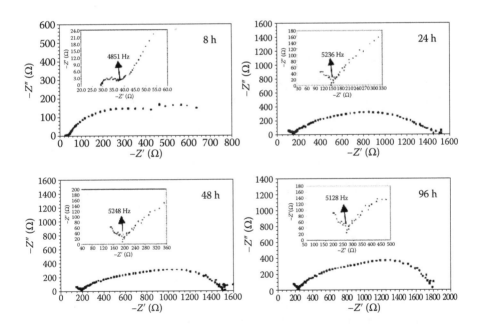

FIGURE 5.44 The Nyquist plot obtained for PPy-coated mild steel electrode in 3.5% NaCl. (Reprinted from *App. Sur. Sci.*, 252, Tuken, T., Yazici, B., and Erbil, M., The corrosion behavior of polypyrrole coating synthesized in phenylphosphonic acid solution, p. 2315, Copyright 2006, with permission from Elsevier.)

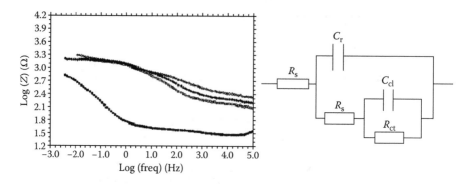

FIGURE 5.45 The Bode plot and circuit obtained for PPy-coated mild steel electrode in 3.5% NaCl: ●, 8 h; ○, 24 h; ▲, 48 h; and △, 96 h. (Reprinted from *App. Sur. Sci.*, 252, Tuken, T., Yazici, B., and Erbil, M., The corrosion behavior of polypyrrole coating synthesized in phenylphosphonic acid solution, p. 2316, Copyright 2006, with permission from Elsevier.)

TABLE 5.4
Impedance, Percentage Efficiency, and Porosity Data Obtained from Figures 5.43 to 5.45

Electrodes	t (h)	E_{corr} (V)	R_p (Ω)	R_{ct} (Ω)	R_f (Ω)	I_{corr} (mA)	$E\%$	$P\%$
MS	4	−0.525	25	25	–	1.040	–	–
MS/PPy	8	−0.380	670	14	656	0.078	96.2	0.055
	24	−0.410	1416	94	1322	0.037	98.2	0.064
	48	−0.440	1473	143	1330	0.035	98.3	0.146
	96	−0.498	1532	192	1340	0.033	98.4	0.752

Source: Reprinted from *App. Sur. Sci.*, 252, Tuken, T., Yazici, B., and Erbil, M., The corrosion behavior of polypyrrole coating synthesized in phenylphosphonic acid solution, p. 2316, Copyright 2006, with permission from Elsevier.

Bereket et al. (2005) carried out electrodeposition of PAN, poly(2-iodoaniline), and poly(aniline-co-2-iodoaniline) on steel surfaces. Figure 5.46 shows Nyquist plot and phase angle–log frequency plot for stainless steel immersed in 0.5 M HCl for 2 h.

A slightly depressed semicircle was observed, and its diameter was used to estimate charge transfer resistance. Nyquist plots for PAN-coated stainless steel at different immersion timed in 0.5 M HCl are shown in Figure 5.47.

The higher value of the polarization resistance obtained in this work was assigned to the protective properties of PAN and was also used to find protection efficiency. Nyquist and phase angle–log frequency plots for poly(2-iodoaniline) (PIAN) and

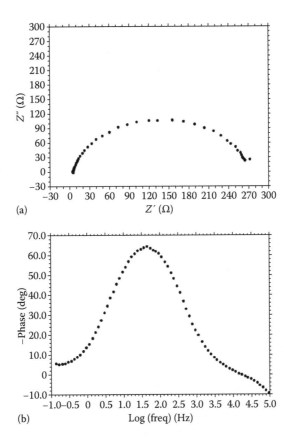

FIGURE 5.46 (a) Nyquist plot for stainless steel, (b) phase angle–log frequency plot for stainless steel in 0.5 M HCl. (Reprinted from *Appl. Surf. Sci.*, 252, Bereket, G., Hur, E., and Sahin, Y., Electrodeposition of polyaniline, poly(2-iodoaniline) and poly(aniline-co-2-iodoaniline) on steel surfaces and corrosion protection of steel, p. 1240, Copyright 2005, with permission from Elsevier.)

poly(aniline-co-2-iodoaniline)(co-PIAN)-coated stainless steel at different immersion times in 0.5 M HCl are shown in Figures 5.48 and 5.49 respectively.

From the examination of polarization resistance, OCP, and percentage efficiency values obtained from EIS measurements, it was concluded that PAN, poly(2-iodoaniline), and poly(aniline-co-2-iodoaniline) coatings exhibited corrosion protection in 0.5 M HCl.

Ozyilmaz et al. (2006) carried out electrochemical synthesis of PAN on stainless steel by applying two scan rates (10 and 50 mV/sec) in 0.1 M aniline containing 0.3 M oxalic acid. The Nyquist and the phase angle–log frequency plots obtained in this work are shown in Figures 5.50 through 5.53.

It was concluded that SS/PAN-L exhibited a barrier effect against corrosion products for longer periods, but this effect decreased significantly in the case of SS/PAN-H after 240 h of immersion. The polymer film obtained at the high scan rate probably had a porous structure than did those obtained at a low scan rate. Therefore,

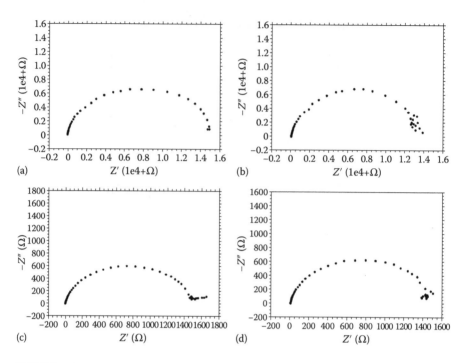

FIGURE 5.47 Nyquist plot for PAN-coated stainless steel at different immersion times: (a) 2 h, (b) 4 h, (c) 24 h, and (d) 48 h in 0.5 M HCl. (Reprinted from *Appl. Surf. Sci.*, 252, Bereket, G., Hur, E., and Sahin, Y., Electrodeposition of polyaniline, poly(2-iodoaniline) and poly(aniline-co-2-iodoaniline) on steel surfaces and corrosion protection of steel, p. 1241, Copyright 2005, with permission from Elsevier.)

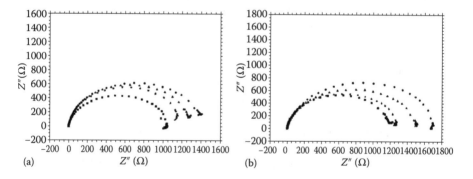

FIGURE 5.48 Nyquist plot for PIAN-coated stainless steel at different immersion time for 2 h, 4 h, 24 h, and 48 h in 0.5 M HCl (a) and Nyquist plot for co-PIAN-coated stainless steel at different immersion time for ●, 2 h; ▲, 4 h; *, 24 h; and ■, 48 h in 0.5 M HCl (b). (Reprinted from *Appl. Surf. Sci.*, 252, Bereket, G., Hur, E., and Sahin, Y., Electrodeposition of polyaniline, poly(2-iodoaniline) and poly(aniline-co-2-iodoaniline) on steel surfaces and corrosion protection of steel, p. 1242, Copyright 2005, with permission from Elsevier.)

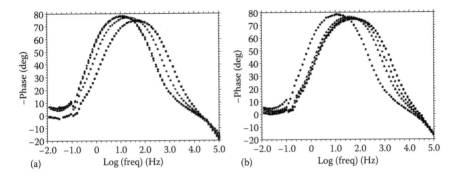

FIGURE 5.49 Phase angle–log frequency plot for PIAN-coated stainless steel at different immersion times for 2 h, 4 h, 24 h, and 48 h in 0.5 M HCl (a) and for co-PIAN-coated stainless steel at different immersion times for ●, 2 h; ▲, 4 h; *, 24 h; and ■, 48 h in 0.5 M HCl (b). (Reprinted from *Appl. Surf. Sci.*, 252, Bereket, G., Hur, E., and Sahin, Y., Electrodeposition of polyaniline, poly(2-iodoaniline) and poly(aniline-co-2-iodoaniline) on steel surfaces and corrosion protection of steel, p. 1242, Copyright 2005, with permission from Elsevier.)

the iron dissolution was intensified toward the metal surface because of the porosity of the film. The catalytic properties of PAN coating under the accelerating effect of chloride ions increased at the interface. The polarization values increased with time. However, the polarization value of SS/PAN-H after 240 h decreased as a result of the saturation of coating with water, and such situation could be related to breakdown of oxide layer and deterioration of the coating.

Yagan et al. (2007) investigated the inhibition of corrosion of mild steel by homopolymer and bilayer coatings of PAN and PPy. Figures 5.54 and 5.55 show impedance spectra of uncoated mild steel electrodes in 0.5 M NaCl and 0.1 HCl.

The Nyquist plots of homolayer and bilayer coatings recorded during long immersion times in NaCl and HCl are shown in Figures 5.56 and 5.57.

Interpretations made in this work using EIS are tabulated in Table 5.5.

Karpagam et al. (2008) studied the corrosion protection of Al2014 T6 alloy by electropolymerized PAN coating, and the impedance results were obtained in the form of Nyquist plots, shown in Figures 5.58 and 5.59.

It was found that PAN coatings on Al2014 T6 alloy synthesized by galvanostatic method with cerium posttreatment exhibited excellent corrosion protection performance in corrosive media containing chloride anions.

Recently, Jadhav and Gelling (2015) prepared titanium dioxide/conducting polymer composite pigments and evaluated their performance against corrosion protection of cold rolled steel in 5% NaCl.

Figures 5.60 and 5.61 show bode plots of coating with 5 wt% tungstate-doped TiO_2/PPy composite, coating with 20 wt% tungstate-doped TiO_2/PPy composite, coating with 5 wt% TiO_2/PAN composite, and coating with 20 wt% TiO_2/PAN composite obtained in this work.

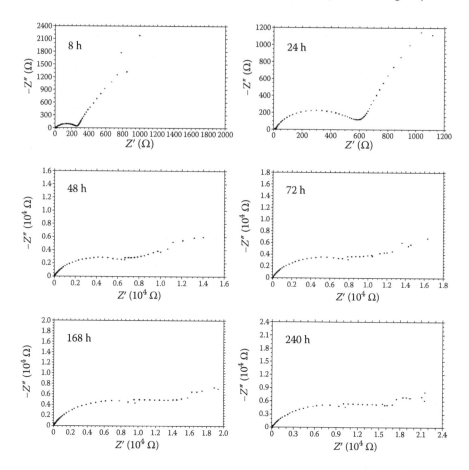

FIGURE 5.50   The Nyquist diagrams obtained for stainless steel (SS)/PAN-L in 0.1 M HCl solution after various exposure times. (Reprinted from *Curr. Appl. Phys.*, 6, Ozyilmaz, A. T., Erbil, M., and Yazici, B., The electrochemical synthesis of polyaniline on stainless steel and its corrosion performance, p. 4, Copyright 2006, with permission from Elsevier.)

Equivalent circuit models used in this work are shown in Figure 5.62.

In the case of coating with 5 wt% tungstate-doped TiO_2/PPy composite, an initial decrease in the coating resistance and a subsequent increase at the 21st day of immersion were noted. It was concluded that the tungstate anions participated in the passivation. In the case of the coating with 5 wt% TiO_2/PAN composite, Warburg impedance appeared at 24 h and the seventh day, indicating the diffusion of species through the coating. At the 14th and 21st days, a second time constant appeared, suggesting a reaction at the metal–coating interface. Only one time constant was employed in case of the coating with 20 wt% TiO_2/PAN composite coating.

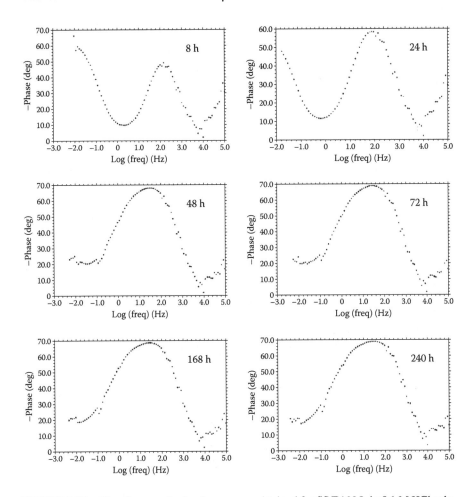

FIGURE 5.51   The phase angle–log freq curves obtained for SS/PAN-L in 0.1 M HCl solution after various exposure times. (Reprinted from *Curr. Appl. Phys.*, 6, Ozyilmaz, A. T., Erbil, M., and Yazici, B., The electrochemical synthesis of polyaniline on stainless steel and its corrosion performance, p. 5, Copyright 2006, with permission from Elsevier.)

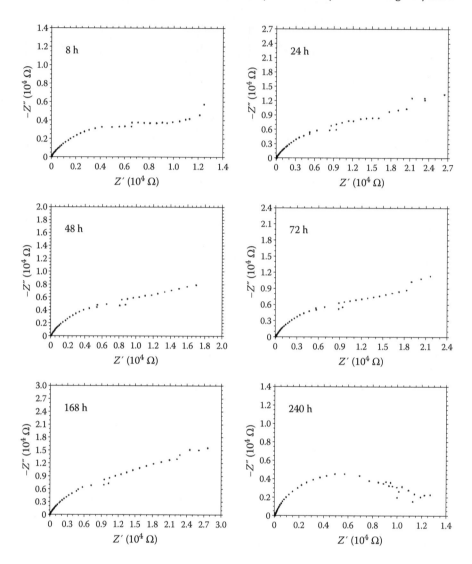

FIGURE 5.52 The Nyquist diagrams obtained for SS/PAN-H in 0.1 M HCl solution after various exposure times. (Reprinted from *Curr. Appl. Phys.*, 6, Ozyilmaz, A. T., Erbil, M., and Yazici, B., The electrochemical synthesis of polyaniline on stainless steel and its corrosion performance, p. 6, Copyright 2006, with permission from Elsevier.)

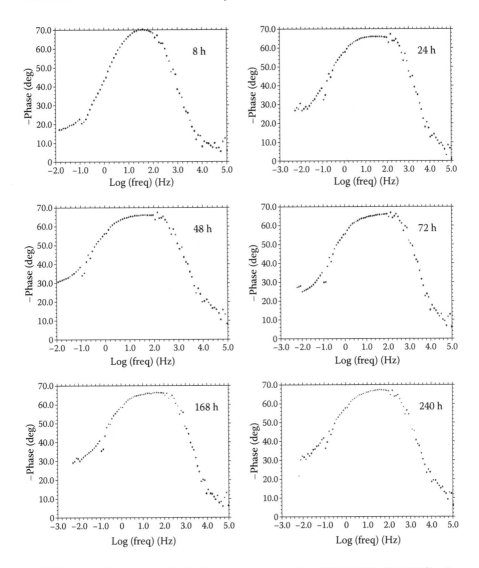

FIGURE 5.53 The phase angle–log freq curves obtained for SS/PAN-H in 0.1 M HCl solution after various exposure times. (Reprinted from *Curr. Appl. Phys.*, 6, Ozyilmaz, A. T., Erbil, M., and Yazici, B., The electrochemical synthesis of polyaniline on stainless steel and its corrosion performance, p. 7, Copyright 2006, with permission from Elsevier.)

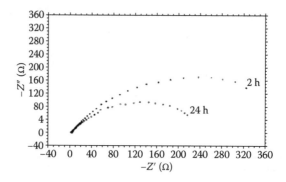

FIGURE 5.54 Impedance spectra of uncoated mild steel electrode in 0.5 M NaCl. (Reprinted from *Prog. Org. Coat.*, 59 Yagan, A, Pekmez, N. O., and Yildiz, A., Inhibition of corrosion of mild steel by homopolymer and bilayer coatings of polyaniline and polypyrrole, p. 300, Copyright 2007, with permission from Elsevier.)

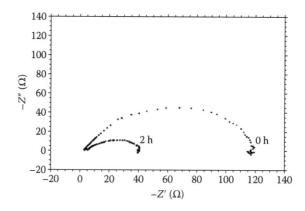

FIGURE 5.55 Impedance spectra of uncoated mild steel electrode in 0.1 M HCl. (Reprinted from *Prog. Org. Coat.*, 59 Yagan, A, Pekmez, N. O., and Yildiz, A., Inhibition of corrosion of mild steel by homopolymer and bilayer coatings of polyaniline and polypyrrole, p. 300, Copyright 2007, with permission from Elsevier.)

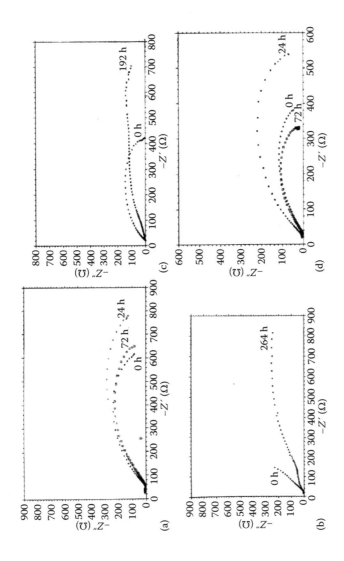

FIGURE 5.56 Impedance spectra of mild steel electrode coated with (a) PAN, (b) PPy, (c) PAN/PPy, and (d) PPy/PAN in 0.5 M NaCl. (Reprinted from *Prog. Organ. Coat.*, 59, Yagan, A., Pekmez, N. O., Yildiz, A., Inhibition of corrosion of mild steel by homopolymer and bilayer coatings of polyaniline and polypyrrole, p. 301, Copyright 2007, with permission from Elsevier.)

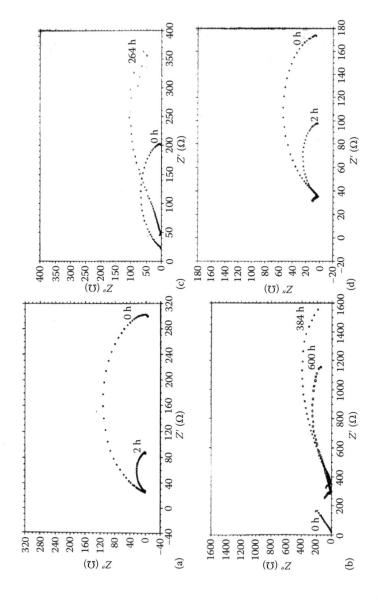

FIGURE 5.57 Impedance spectra of mild steel electrode coated with (a) PAN, (b) PPy, (c) PAN/PPy, and (d) PPy/PAN in 0.1 M HCl. (Reprinted from *Prog. Organ. Coat.*, 59, Yagan, A., Pekmez, N. O., Yildiz, A., Inhibition of corrosion of mild steel by homopolymer and bilayer coatings of polyaniline and polypyrrole, p. 301, Copyright 2007, with permission from Elsevier.)

TABLE 5.5
Impedance Analysis

Coating	EIS Interpretations
PAN and PPY/PAN	Decay of protection after immersion of 24 h as a result of penetration of corrosive chloride ions
PPy as compared with PAN/PPy	Increased charge transfer resistance during long immersion
PAN and PPy/PAN as compared with PAN/PPy	Less protection and decay of protection after 2 h of immersion in HCl
PPy	In the beginning of immersion, Warburg behavior indicated the resistance of coating against the diffusion of corrosive ions Maximum protection

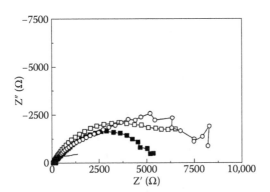

FIGURE 5.58 Impedance behavior of uncoated and galvanostatic electropolymerized PAN-coated Al2014 T6 alloy in 1% NaCl: —, blank; □, 15 mA; ■, 20 mA; o, 25 mA. (Reprinted from *Curr. Appl. Phys.*, 8, Karpagam, V., Sathiyanarayanan, S., and Venkatachari, G., Studies on corrosion protection of Al2014 T6 alloy by electropolymerized polyaniline coating, p. 96, Copyright 2008, with permission from Elsevier.)

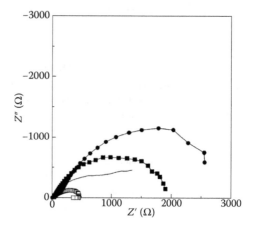

FIGURE 5.59 Impedance behavior of potentiostatically electropolymerized PAN-coated Al2014 T6 alloy in 1% NaCl: —, blank; O, 1.7 V; Π, 1.8 V; Π, 1.9 V. (Reprinted from *Curr. Appl. Phys.*, 8, Karpagam, V., Sathiyanarayanan, S., and Venkatachari, G., Studies on corrosion protection of Al2014 T6 alloy by electropolymerized polyaniline coating, p. 98, Copyright 2008, with permission from Elsevier.)

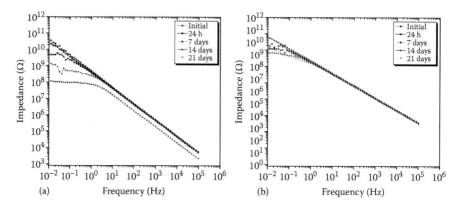

FIGURE 5.60 Bode plot of Ti PPy W 5 (a) and Ti PPy W20 (b). (With kind permission from Springer Science+Business Media: *J. Coat. Technol. Res.* Titanium dioxide/conducting polymers composite pigments for corrosion protection of cold rolled steel, 12(1), p. 148, 2015, Jadhav, N. G. and Gelling, V. J. © American Coatings Association 2014.)

155

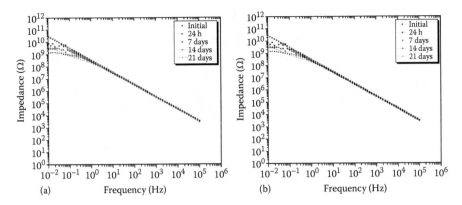

(a)   Frequency (Hz)   (b)   Frequency (Hz)

FIGURE 5.61 Bode plot of Ti PAN15 (a) and Ti PAN20 (b). (With kind permission from Springer Science+Business Media: *J. Coat. Technol. Res.* Titanium dioxide/conducting polymers composite pigments for corrosion protection of cold rolled steel, 12(1), p. 148, 2015, Jadhav, N. G. and Gelling, V. J. © American Coatings Association 2014.)

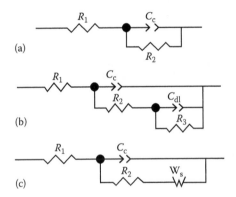

FIGURE 5.62 (a) Use of Randles circuit: R_1 is solution and R_2 is coating resistance and C_c coating capacitance. (b) Use of second time constant: C_{dl} is double layer capacitance. (c) Use of Warburg impedance, W_s. (With kind permission from Springer Science+Business Media: *J. Coat. Technol. Res.* Titanium dioxide/conducting polymers composite pigments for corrosion protection of cold rolled steel, 12(1), p. 149, 2015, Jadhav, N. G. and Gelling, V. J. © American Coatings Association 2014.)

5.5　ZERO-RESISTANCE AMMETRY

Zero-resistance ammetry is used for determining corrosion rates in couples arising from different metals in contact or in corrosion cell created by local variations in electrochemical conditions. The zero-resistance ammeter is based on an operational amplifier that can measure current flow by using the output current of the amplifier to offset the input current from the corrosion cell. As a result, there is no net passage of current from the system whist the feedback current of the amplifier will be equal to that in the corrosion cell and can be measured without interference with the corrosion reactions in the galvanic cell (Von Fraunhofer & Boxall, 1976). The circuit diagram for zero-resistance ammeter is shown in Figure 5.63.

Deshpande et al. (2013) recently investigated galvanic corrosion resistance of conducting poly(*o*-anisidine)-coated low-carbon steel samples by using zero-resistance ammeter. A plot of the galvanic current recorded during this work is shown in Figure 5.64.

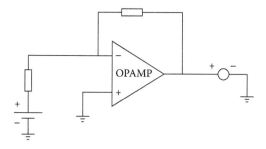

FIGURE 5.63　Zero-resistance ammeter—circuit diagram. (From Deshpande, P. et al., *U.P.B. Sci., Series B*, 75, p. 253, 2013. With permission.)

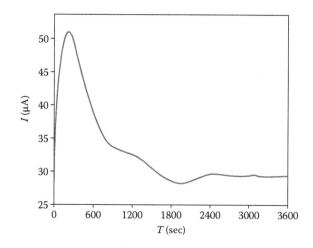

FIGURE 5.64　Plot of the galvanic current. (From Deshpande, P. et al., *U.P.B. Sci., Series B*, 75, p. 253, 2013. With permission.)

An increase in galvanic current was observed until 51 μA during a 4 min period. Subsequently, a constant decrement was observed for 20 min, which in turn stabilizes up to 28 μA. The authors argued that since conducting polymer exhibited very less galvanic current compared with conventional paint-coated samples, conducting polymer coating protects steel samples from galvanic corrosion better than conventional painted samples do.

REFERENCES

Al Dulami, A. A., Hashim, S., & Khan, M. I. (2011). Corrosion protection of carbon steel using polyaniline composite with inorganic pigments. *Sains Malays.*, *40*, 762.

Barnartt, S. (1978). Electrochemical nature of corrosion. *Electrochemical Techniques for Corrosion* (p. 1). Texas: NACE.

Bereket, G., Hur, E., & Sahin, Y. (2005). Electrodeposition of polyaniline, poly(2-iodoaniline), and poly(aniline-co-2-iodoaniline) on steel surfaces and corrosion protection of steel. *Appl. Surf. Sci.*, *252*, 1233.

Bernard, M., Deslouis, C., EI Moustafid, T., Hugot-Legoff, A., Joiret, S., & Tribollet, B. (1999). Combined impedance and raman analysis in the study of corrosion protection of iron by polyaniline. *Synth. Met.*, *102*, 1382.

Bockris, J. O'M., & Reddy, A. K. N. (2000). *Modern Electrochemistry*, 2nd ed., vol. 2B (p. 1661). New York: Kluwer Academic Publishers.

Brasher, D., & Kingbury, A. (1954). Electrical measurements in the study of immersed paint coatings on metal. I: Comparision between capacitance and gravimetric methods of estimating water uptake. *J. Appl. Chem.*, *4*, 62.

Chaudhari, S., Sainkar, S. R., & Patil, P. P. (2007). Poly(o-ethylanine) coatings for stainless steel protection. *Prog. Org. Coat.*, *58*, 60.

Chaudhari, S., Gaikwad, A. B., & Patil, P. P. (2010). *Coat. Technol. Res.*, *7*, 119.

Colreavy, J., & Scantlebury, J. (1995). Electrochemical impedance spectroscopy to monitor the influence of surface preparation on the corrosion characteristics of mild steel MAG welds. *J. Mater. Process. Technol.*, *55*, 206.

Deshpande, P., Deshpande, P., & Vagge, S. (2013). Galvanic corrosion investigations on conducting poly(o-anisidine) coated low carbon steel samples by using zero resistance ammeter. *U.P.B. Sci., Ser. B*, *75*, 257.

Jadhav, N., & Gelling, V. (2015). Titanium dioxide/conducting polymers composite pigments for corrosion protection of cold rolled steel. *J. Coat. Technol. Res.*, *12*, 137.

Karpagam, V., Sathiyanarayanan, S., & Venkatachari, G. (2008). Studies on corrosion protection of Al 2024 T 6 alloy by electropolymerized polyanline coating. *Curr. Appl. Phys.*, *8*, 93.

Kelly, R., & Scully, J. (2003). *ASM Hand Book, vol. 13A, Corrosion: Fundamentals, Testing and Protection* (p. 69). Materials Park, OH: ASM International.

Kelly, R., Scully, J., Shoesmith, D., & Buchneit, R. (2003). *Electrochemical Techniques in Corrosion Science and Engineering* (p. 54). New York: Marcel Dekker, Inc.

Kinlen, P., Silverman, D., & Jeffreys, C. (1997). Corrosion protection using polyaniline coating formulation. *Synth. Met.*, *85*, 1327.

Li, P., Tan, T., & Lee, J. (1997). Corrosion protection of mild steel by electroactive polyaniline coatings. *Synth. Met.*, *88*, 241.

Lu, W., Elsenbaumer, R., & Wessling, B. (1995). Corrosion protection of mild steel by coatings containing polyaniline. *Synth. Met.*, *71*, 2165.

Ozyilmaz, A., Erbil, M., & Yazici, B. (2006). The electrochemical synthesis of polyaniline on stainless steel and its corrosion performance. *Curr. Appl. Phys.*, *6*, 1.

Radhakrishnan, S., Siju, C., Mahanta, D., Patil, S., & Madras, G. (2009). Conducting poly-aniline-nano-TiO$_2$ composite for smart corrosion resistant coatings. *Electrochim. Acta, 54*, 1251.

Rajagopalan, R., & Iroh, J. (2002). Corrosion performance of polyaniline-polypyrrole composite coatings applied to low carbon steel. *Surface Eng., 18*, 62.

Rout, T., Jha, G., Singh, A., Bandyopadhyay, N., & Mohanty, O. (2003). Development of conducting polyaniline coating: A novel apporach to superior corrosion resistance. *Surf. Coat. Technol., 167*, 15.

Samui, A., Patankar, A., Rangarajan, J., & Deb, P. (2003). Study of polyaniline containing paint for corrosion protection. *Prog, Org. Coat., 47*, 5.

Sathiyanarayanan, S., Muthukrishnan, S., Venkatachari, G., & Trivedi, D. (2005). Corrosion protection of steel by polyaniline (PAN) pigmented paint coating. *Prog. Org. Coat., 53*, 299.

Sazou, D., & Georgolios, C. (1997). Formation of conducting polyaniline coatings on iron surfaces by electropolymerization of aniline in aqeous solutions. *J. Electroanal. Chem., 429*, 81.

Stern, M., & Geary, A. (1957). A theroretical analysis of the shape of polarization curves. *J. Electrochem. Soc., 104*(1), 56.

Syed Azim, S., Sathiyanarayanan, S., & Venkatachari, G. (2006). Anti corrosive properties of PAN-ATMP polymer containing organic coatings. *Prog. Org. Coat., 56*, 157.

Tallman, D., Spinks, G., Dominis, A., & Wallace, G. (2002). Electroactive conducting polymers for corrosion control: Part 2. Ferrous metals. *J. Solid State Electrochem., 6*, 85.

Tuken, T., Yazici, B., & Erbil, M. (2006). The corrosion behaviour of polypyrrole coating synthesized in phenylphosphonic acid solution. *Appl. Surf. Sci., 252*, 2341.

Von Fraunhofer, J., & Boxall, J. (1976). The protective action of paint film. *Surf. Technol., 4*, 187.

Wei, Y., Wang, J., Jia, X., Yeh, J., & Spellane, P. (1995). Polyaniline as corrosion protection coatings on cold rolled steel. *Polymer, 36*(3), 4536.

Yagan, A., Pekmez, N. O., & Yildiz, A. (2007). Inhibition of corrosion of mild steel by homo-polymer and bilayer coatings of polyaniline and polypyrrole. *Prog. Org. Coat., 59*, 297.

6 Strategies for Improving the Protective Efficiency of Coatings Based on Conducting Polymers

6.1 INTRODUCTION

Extensive research and field studies have resulted so far in novel approaches by which conducting polymer (CP)-based coatings seem promising in replacing effectively the chromate-based processes because of the favorable self-healing property of CPs. Through the implementation of novel approaches and evaluation results, a deeper understanding of how corrosion protection by CPs works is simultaneously gained. In view of Chapter 3, corrosion protection by CPs can be understood so far by several mechanisms: (1) active electronic barrier, (2) anodic protection and self-healing either through "ennobling effect" or controlled inhibitor release after an electrochemical triggering due to corrosion initiation, (3) cathodic protection through a displacement of the oxygen reduction reaction (ORR) from the metal–electrolyte interface to the CP–electrolyte interface, and (4) physical barrier between metal and environment. From previous chapters, it became clear that more than one mechanism can work simultaneously providing anticorrosion protection. Therefore, the success or failure of an organic coating depends on a variety of factors, usually more than one. Although much more research is certainly needed to understand the mechanisms by which CP-based coatings can prevent or retard corrosion of metals and alloys, the knowledge acquired so far by numerous studies may lead researchers to properly design coatings based on CPs by avoiding their negative effects and improving their protective efficiency. Many approaches for designing effective CP-based coatings can be found in Chapter 4 among different coating formulations, all aiming both to improve the anticorrosion performance and prolong the lifetime of coatings. These include different top coats of CP primers, CP-based blends, composites, and paints with CP as additives.

This chapter outlines, first, in Section 6.2, several problems encountered when practical use of CP-based coatings is required in severe corrosive environments. Respective strategies implemented to improve the functionality and protective efficiency of CP films deposited on active metal substrates are outlined. Sections 6.3 and 6.4 delineate strategies that came out as a result of the knowledge gradually gained and classified/unified under the term protection mechanisms by which polyaniline

(PAN), polypyrrole (PPy), and polythiophene (PTh) are considered to perform as anticorrosion pigments under different corrosion conditions. Emphasis is placed on the active role of CPs, which enables them to function as intelligent anticorrosive materials by exploiting their electrochemical activity and self-healing property.

6.2 PROBLEMS ENCOUNTERED WHEN USING CONDUCTING POLYMERS IN ANTICORROSION TECHNOLOGY

From the large number of studies and reviews devoted to CP-based organic coatings and previous chapters of the present book, it becomes clear that CPs can provide protection against various forms of metal corrosion, mostly against uniform or generalized corrosion. It becomes also certain that in the presence of relatively large defects as in the case of localized (pitting or crevice) corrosion, CPs frequently fail and protection mechanisms 1–4 all become ineffective, even though they may work perfectly for small ones. In general, the major problems encountered in utilizing CP-based coatings to inhibit corrosion can be summarized as follows:

- Weak and short-term adhesion between the coating and substrates
- Delamination and blistering
- Dopants possibly activating corrosion as a result of ion exchange between the doping anion and the ions present in the surrounding environment
- Insertion of water or electrolyte in the coatings
- CP slow reoxidation of its reduced form and delay in recovering its active form
- CPs in their oxidized state lacking stability
- CP processability and difficulty to incorporate into the conventional coating systems

6.2.1 ADHESION

Weak adhesion is the main problem in most of the cases where failure of CP-based coatings is observed. The interface between an oxidizable metal and the CP (M/CP) is one of the major issues to be understood as maybe being responsible for the blistering, breakdown, and delamination of the protective CP-based coating. The M/CP is an ill-defined interface between two different conducting materials separated perhaps by a semiconducting oxide. Strong and long-term adhesion is an ultimate goal that could better be achieved by covalent bonds between CPs and metal substrate. When CPs are deposited through electrochemical or chemical polymerization of monomers on metal surfaces, covalent bonding is not possible. However, PAN, PPy, and PTh in their oxidized state are usually strongly adherent on the surfaces of active metals depending on the polymer film thickness and roughness of the substrate. This strong adhesion is attributed mainly to electrostatic and weak chemical interactions. CPs are generally poorly adherent in their reduced state. Therefore, when a pinhole is generated on the coating and reduction of the polymer starts, then delamination may occur around the pinhole if the self-healing capability of the coating is not fast enough to reoxidize the CP and repair the oxide at the metal active area before its spreading. Therefore, adhesion is of major importance and a number of factors should be controlled before a CP-based protective coating is applied on the surface of an active metal.

A common and simple practice to ensure good adhesion and the actual performance of CP-based coatings is to use a binder such as epoxy, acrylic, or polyurethane (Elhalawany et al., 2014). This type of CP-based formulation provides, in certain cases, superior performance than electrodeposited CP coatings do. However, a drawback of this strategy is lesser contact between CP and the metal substrate.

Pretreatment of the surface of active metals before polymer deposition is also an efficient practice. For example, in the case of pyrrole (Py) electropolymerization on oxidizable metals, strongly adherent PPy films could be deposited on Fe or mild steel without an induction period when the substrate was pretreated with dilute nitric acid (Ferreira et al., 1996). Reut et al. (1999) compared mechanical and chemical in 10% HNO_3 pretreatment techniques before the control potential sweep deposition of Py on Fe surfaces. A greater shift of the open circuit potential to positive values was observed for the chemically treated PPy-coated Fe surface assigned to the passivation of Fe facilitated by the formation of iron nitrides during chemical treatment. Moreover, electrochemical polymerization of aniline (AN) resulted in the deposition of uniform, compact, and strongly adherent PAN coatings on nickel-plated Cu surfaces. Enhancement of the barrier properties of the PAN film on Cu/Ni surfaces is caused by the formation of freshly produced nickel oxide layers and the reduction of the polymer film at Ni–PAN (Ozyilmaz, 2006).

Several other strategies have been used to improve adhesion of CP on oxidizable metals (Lacaze et al., 2010), such as the following:

- Bilayer/multilayer coatings, composed of at least two types of CP deposited electrochemically. In these cases, the first layer enhances adhesion to the metal, although no covalent bonds between the metal and the polymer exist.
- Bifunctional coupling agents capable of chemically reacting with the metal and with the polymer. Such coupling reagents comprise two functional groups, one like siloxane, thiol, or phosphonate, which will strongly react with metal hydroxides or oxides, and the other, an unsaturated group or a monomer, which will react with the polymer. Such adhesion primer layers can consist of organic siloxanes on top of which the CP could be deposited (Najari et al., 2009).
- Thiol-based pretreatments, which seem promising for specific applications but also require quite a long time to completely cover the metal surface through the formation of a well-defined monolayer (Jaehne et al., 2008).
- Electroreduction of diazonium salts on oxidizable metals that allows grafting of thin layers of organic compounds covalently attached at the metal surface, generally based on oligophenylene chains. This reaction, first applied on a carbon substrate (Allonque et al., 1997), can be applied on several oxidizable metals (Fe, mild steel, Ni, Co, Al) (Adenier et al., 2002). Such layers exhibited anticorrosion properties and are used as primary layers for surface polymerization of

monomers leading to non-CPs (Santos et al., 2008). A drawback of diazonium salts is that they are nonconducting and hence do not allow electrochemical deposition of a second material. This can be overcome when the AN dimer, 4-aminodiphenylamine, is used as the starting diazonium salt (Santos et al., 2008). In that case, the grafted layer retains electroactivity and AN oxidation, and in turn, polymerization may occur, resulting in a strongly adherent PAN film on the modified metal surface.

6.2.2 DELAMINATION AND BLISTERING

When a small scratch in the coating occurs, reduction of the polymer, which is supposed to induce self-healing, causes delamination around the scratch. As a consequence, the scratch size increases and the steady state existing between the metal and the CP coating is perturbed, as further reduction of the polymer is needed to repair the new active area in which there is direct contact between the bare metal and the environment. Eventually, this will lead to a complete failure of the coating and to corrosion extending from the initial scratch to large areas of the metal.

Blistering is one of the forms of failure in CP-based coatings. It occurs when the chloride content of fluid in blisters formed on seawater immersion is lower than that of the seawater. The substrate surface under blisters is free of corrosion, but creation and expansion of blisters lead gradually to poor interfacial adhesion of the coating to the metal substrate. Since the concentration gradient is the driving force for blistering, the ion-barrier properties of CP coatings are critical for an optimized protective efficiency. For example, in the case of PAN dispersed in a cationic polymer matrix, blistering was attributed to local cation-selective areas of PAN coating due to cationic defects that allow the transport of cations (Wang et al., 2007a,b).

6.2.3 DOPANT ION EXCHANGE PROPERTIES

A variety of organic and inorganic dopants of different size and chemistry can be used during the chemical or electrochemical synthesis of CPs. The size of the doping anion directly influences the corrosion performance of CPs as CPs may act as anion permselective membranes exchanging the doping anions with external aggressive anions such as chlorides and bromides. It was shown that PAN and poly(aniline-co-o-aminophenol) formed within the Nafion membrane on stainless steel (SS) of AISI304 type provides an almost complete protection of the substrate against localized corrosion in chloride-containing acid and neutral solutions, preventing chlorides to reach the substrate (Kosseoglou et al., 2011; Sazou & Kosseoglou, 2006, 2009). This effect can be understood by considering (1) the cationic permselectivity of Nafion that prevents chloride transport into the composite layer and (2) the contribution of proton transport than anions in charge compensation processes due to the fact that sulfonate groups of Nafion function as inner dopants. Moreover, in the case of PPy in alkaline media, it was shown that PPy doped with large counter anions shows better corrosion resistance than does PPy doped with small counter ions (Qi et al., 2012).

Identification of the crucial role of the dopant (X^-) in the performance of CPs as corrosion protection coatings led to approaches exploiting how the molecular structure and other properties of dopants could rather add to the protective efficiency of CP-based coating than being a source of problems. The chemical nature and size of the dopant may determine (1) the solution solubility and, hence, the processability of the CP, (2) the redox and chemical properties, and (3) the structure of the resultant coating.

1. The formation of an impervious CP-based layer using conventional coating technologies depends mainly on the ability to form polymer solutions or stable dispersions. The solubility of PPy, PAN, and their derivatives is dopant dependent. The effect of sulfonic dopants on the solublility and structure of PAN and its derivative was systematically investigated (Huang & Wan, 1999). Sulfonation renders PAN soluble in common organic solvents such as xylene, CH_2Cl_2, and $CHCl_3$. Moreover, doping with phosphoric acid diesters found to render PANs both melt and solution processable (Paul & Pillai, 2001).

2. The possible role of metal-dopant salts in interfacial oxide growth and substrate passivation has been investigated (Kinlen et al., 1999, 2002; Souza 2007). There are experimental evidences that when galvanic coupling takes place at the M/CP interface system, the CP is reduced, resulting in the expulsion of dopant anions (Barisci et al., 1998). In the case of PAN, dopant release is triggered by a pH increase due to the ORR at the CP (Kendig et al., 2003). The release of inhibitor anions existing as dopants in the CP matrix for charge compensation is a promising way to construct intelligent inhibitor delivery systems. Perfect performance of these systems required that the intelligent CP-based coating would release inhibitors when corrosion occurs and, within a very short time, stop the release since the corrosion stops because of the inhibiting action of the released inhibitor. In the case of PPy, the release of dopant ion is caused by the electrochemical reduction of the polymer itself at a defect site where corrosion proceeds (Paliwoda-Porebska et al., 2005, 2006). Although the associated interfacial potential changes in the case of PPy might be considered as a precise way for intelligent release of an inhibiting anion, the possible incorporation of cations might lead to complications. Indeed, electrochemical quartz microbalance (EQCM) studies combined with cyclic voltammetry have shown that the mobility of the dopant ions during the reduction/oxidation of the CP depends on the mass, volume and valency of the dopant ion (Khalkhali et al., 2003; Koehler et al., 2007; Maia et al., 1996). These properties of ions involved in doping, as well as the insertion/expulsion of the solvent, control the success or failure of the inhibitor release.

3. Studies on the anticorrosion performance of PAN in conjunction with its anion exchange nature showed that PAN in both its doped, emeraldine salt (ES) and undoped, emeraldine base (EB) states need to be covered with a cationic membrane to prevent undercoating corrosion of cold rolled steel

(CRS) surfaces in 3.5 wt% NaCl solutions and delay delamination (Wang et al., 2007b). The behavior of ideal anionic, cationic, and bipolar PAN coatings can be understood by the scheme depicted in Figure 6.1. In the case of PAN primer (EB state) alone (Figure 6.1a), the OH^- ions migrate from the coating side to the bulk, while Cl^- ions migrate from the bulk solution into the coating/metal interface. The OH^- ions are generated cathodically because of the ORR. If the PAN is in its ES state, doping ions might be also exchanged with Cl^- ions. This is because of a potential gradient, which results in ion transport in reverse directions. For the same reason, Na^+ and Fe^{2+} ions migrate in reverse directions in the case of the cationic membrane (Figure 6.2b), whereas neither Cl^- ions can insert to nor OH^- ions can leave the coating–metal interface. The OH^- ions trapped across the coating–metal interface lead to a pH increase and the formation of iron oxide. It is suggested that delamination occurs when the concentration of sodium hydroxide reaches a certain value. In the case of the ideal bipolar

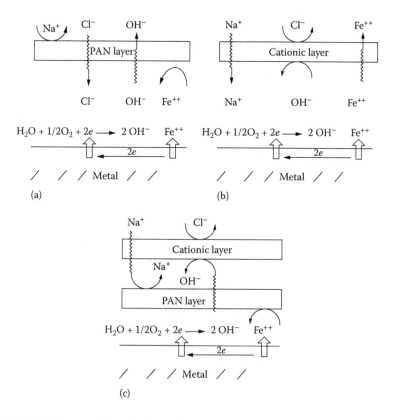

FIGURE 6.1 Scheme of how ionic coatings can act when immersed in NaCl solution: (a) ideal PAN anionic membrane, (b) ideal cationic coating, and (c) ideal bipolar coating. (From Wang, J., Torardi, C. C., Duch, M. W., *Synth. Met.*, *157*, 851–858, 2007.)

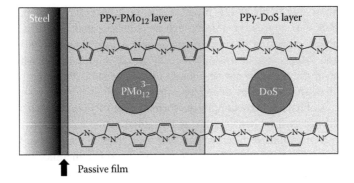

FIGURE 6.2 **(See color insert.)** Bilayer PPy film designed to stabilize the passive film on the steel by the action of phosphomolybdate ions in the inner PPy layer, and to control the ion perm-selectivity by bulky dopants (dodecylsulfates) in the outer PPy layer. (From Ohtsuka, T., *Int. J. Corros.*, Volume 2012, Article ID 915090, 7 pp., 2012.)

PAN coating, the chloride-induced metal dissolution is inhibited since the cationic top coat prevents Cl⁻ penetration and the OH⁻ ions (or doping ions) are kept inside the coating/metal interface.

The ion exchange capacity of PAN depends on the polymerization conditions and the doping or protonation degree as well as on the chemical nature of the dopant (Gospodinova & Terlemezyan, 1998; Nagaoka et al., 1997).

To sum up, the effect of dopant ion is very important when CPs are utilized as primer without a top coat. In general, dopant ion may be exchanged with solution chloride ions, and dopant ion may induce delamination from large defect sites. As it was shown for PPy and PAN, the delamination rate is determined by the reduction kinetics of the polymer (Rohwerder, 2009).

6.2.4 Delay in Recovering CP in Its Active Form

The barrier effect and the oxidative property of CPs induce anodic protection. Eventually, as was shown, the M/CP system acquires a potential that corresponds to the CP layer and depends on the oxidation state of the CP. This oxidation state is determined by the doping level and, hence, the conductivity of the CP. For example:

$$PPy^{n+} + \frac{n}{x}A^{x-} + me \rightleftharpoons PPy^{(n-m)+} \cdot \frac{(n-m)}{x}A^{x-} + \frac{m}{x}A_{(aq)}^{x-}P \qquad (6.1)$$

The oxidation degree and the conductivity of CPs have been found to decline with longer exposure to corrosive environment containing chlorides. If oxidants in the environment (typically oxygen) reoxidize the degraded CP layer, the oxidation degree and conductivity can be recovered. Therefore, the oxidative power of the CP can be prolonged and the passive state of the steel underneath the PPy layer can be kept for a longer period.

A problem arising in the anodic protection is the breakdown of the passive oxide due to the attack of aggressive anions such as chlorides and bromides, leading to localized corrosion (pitting or crevice corrosion). A local breakdown is often followed by large defects, and corrosion proceeds at such high rates for which the oxidative power of the CP may not be enough to stop or slow down the metal active dissolution. It is shown that by choosing a proper anion as dopant, a stable passive state may be established. For example, the formation of PPy coatings on steel oxalate ions has been widely used because during the initial active dissolution, oxalate ions formed with iron cations ferrous oxalate that passivates the substrate. In a second step, the PPy-oxalate coating is deposited. It exhibits good adherence and protective efficiency against the corrosion of steels in chloride-containing media. However, this protection did not continue for a long period.

Alternative approaches to prolong the protective action of the CP coating include the following:

a. An additional overcoat consisting of PPy doped with organic ions of a large size has been deposited on the PPy-oxalate (Ohtsuka, 2012). The dodecyl-sulfate, $C_{12}H_{25}OSO_3^-$ (DoS), ion used as an immobile dopant in the PPy can prevent penetration of aggressive ions such as chloride ions. The PPy-DoS coating is considered as a membrane with negatively charged fixed sites and, hence, with cationic permselectivity (Figure 6.2).

b. Incorporation of oxide particles or metal particles has been introduced into the PPy-oxalate to enhance the corrosion resistivity (Ohtsuka, 2012).

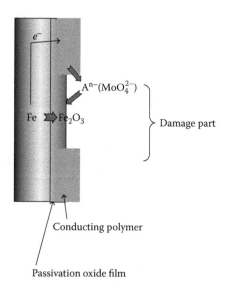

FIGURE 6.3 Model for self-healing property of PPy-PMo-DoS bilayered PPy. Reformation of the passive oxide film is possible by molybdate anions release from PPy film. (From Ohtsuka, T., *Int. J. Corros.*, Volume 2012, Article ID 915090, 7 pp., 2012.)

c. Use of a bilayer PPy, which combines the stabilization of the inner passive layer and the prevention of the insertion of aggressive ions (Figure 6.3). Thus, the steel was covered first by PPy doped with phosphomolybdate ions, $PMo_{12}O_{40}^{+}$ (PMo), and then by PPy-DoS. As was shown by EQCM for PPy-PMo in comparison with the bilayer PPy-PMo/PPy-DoS coating deposited on a gold electrode, the latter coating acts as a cationic permse-lective membrane. It seems that the PPy-PMo/PPy-DoS coating, 5 μm in thickness, renders steel in its passive state for longer time (190 h) than the PPy-DoS coating does (10 h) when the coated steel is immersed in 3.5 wt% NaCl solution (Kowalski et al., 2007).

The protective performance of the well-designed bilayer PPy-PMo/PPy-DoS coating was further examined in the presence of a small flow inserted by a cutting knife after 2 h of immersion in 3.5 wt% NaCl solution. It was found that the open circuit potential (E_{OC}) temporarily decreases. However, because of the self-healing property of the bilayer coating, the E_{OC} recovers in values corresponding to the passive state. Molybdate salt was detected in the defect site by Raman scattering spectroscopy, suggesting that molybdate ions react with ferric ions on the flawed site to produce the ferric molybdate film.

Though the bilayer PPy-PMo/PPy-DoS coating shows optimized properties in terms of barrier effect, the presence of doping ion that enhances passivity and the existence of additional bulky doping ion at the outermost layer that prevents the penetration of aggressive ions, decline of its anticorrosive ability seems unavoidable during prolonged exposition of the coated metal in chloride-containing media.

6.3 CP COATINGS WITH IMPROVED PROTECTIVE PROPERTIES

To overcome many of the CP-based coatings limitations, it should be ensured that charge stored in the polymer system and consumed to oxidize substrate metal to produce passive layer could be regenerated for the polymer to retain its protective properties. Additionally, the processability, adherence, porosity, and exchange properties of CPs could be improved in view of designing a long-term effective CP-based coating. Several approaches designed to improve CPs' protective properties and functionality, such as copolymerization, especially of AN and its derivatives, were used to improve mainly solubility and processability problems of the PAN and improve homogeneous dispersion of PAN in blends and, in particular, in conventional paint coatings containing PAN additives.

6.3.1 COPOLYMERS

Structural modification of the CP backbone by copolymerization affects various properties of the polymer, such as conductivity, porosity, adherence to the substrate, and stability. In particular, in the case of PPy, the water uptaking property is one of the main drawbacks in the use of PPy as protective coating. Poly(N-methyl pyrrole) and its copolymer with pyrrole showed improved protective properties for mild steel protection (Tuken et al., 2007). A series of poly(pyrrole-co-N-methyl pyrrole) copolymers electrodeposited on Cu from polymerization solutions containing different ratios of monomers. Optimum protective performance of the Cu surfaces in 3.5 wt% NaCl was found for 8:2 ratio of pyrrole/N-methyl pyrrole monomers (Çakmakcı et al., 2013). The improved anticorrosion efficiency of this coating was attributed to the highest interaction between the copolymer and the Cu substrate. This was supported by theoretical calculations indicating that for the 8:2 ratio, the copolymer has the most linear structure and, hence, the best interaction (electron donor) with Cu.

Py was copolymerized electrochemically with substituted ANs forming adherent films on metal substrate. For instance, ter-polymer poly(pyrrole–co-o-anisidine–co-toluidine) was electrochemically deposited on the surface of low-carbon steel (Yalcinkaya et al., 2010). This ter-polymer exhibited improved resistance to water permeation. The copolymer poly(pyrrole–co-o-toluidine) was synthesized electrochemically on mild steel only at low temperatures, exhibiting the best protective efficiency in 3.5 wt% NaCl for a pyrrole:o-toluidine feed 8:2 ratio (Yalcinkaya et al., 2008). PPy and poly(pyrrole–co-o-anisidine) were electrochemically synthesized on the surface of 3102 aluminum alloy (Mert & Yazici, 2011). This copolymer demonstrated better protection for the aluminum alloy than the simple PPy coatings, which was attributed to the presence of the hydrophobic methoxy group that prevents water permeation.

On the other hand, evidence exists that hydrophilic groups attached to the PAN chain lower the anticorrosive performance of PAN-based coatings. For instance, poly(aniline-co-metanilic) copolymers prepared chemically and deposited on carbon steel (Xing et al., 2014). These copolymers provide lesser protection of the coated carbon steel in 1 M H_2SO_4 than the PAN itself, and the lower the sulfur content in the copolymer, the better is the protective ability. The hydrophilic character of the sulfonic group seems unfavorable for anticorrosion performance of the coating. Nevertheless, other studies show that copolymers of PAN with its derivatives containing sulfonic, carboxylic, and methoxy groups exhibit improved protection efficiency owing to the modification of polymer backbone that affects adherence, compactness, porosity and the doping/dedoping process of the coating. Such sulfonic copolymers include:

- Poly(aniline-co-amino-naphthol-sulphonic acid) nanowire coating synthesized electrochemically on the surface of iron (Bhandari et al., 2010). Increased corrosion resistance in comparison with a PAN coating was attributed to the strong adherence of the copolymer due to the interaction of side groups to the iron substrate as well as its superior adsorption to the iron substrate.
- Poly(aniline-co-4-amino-3-hydroxy-naphthalene-1-sulfonic acid) synthesized by chemical oxidative polymerization. In this synthesis, 4-amino-3-hydroxy-naphthalene-1-sulfonic acid was a dopant. The corrosion inhibiting properties of the copolymer were studied on an iron substrate in 1 M HCl. It was found that as the concentration of the inhibitor dopant increased in the monomer feed, the corrosion inhibition efficiency also increased (Bhandari et al., 2011).

- Poly(aniline-co-m-amino benzoic acid) was electropolymerized on the surface of steel (Kamaraj et al., 2010). It was found that the improved corrosion protection ability of the synthesized copolymer was attributed to the better compactness of the resulting films.

Moreover, self-doped conducting PAN films were deposited on passive SS of type AISI 304 substrates by cyclic potential sweep deposition from aqueous mixed monomer solutions of AN and different aminobenzenesulfonic acids (ABSAs) (2-, 3-, or 4-ABSA), without the presence of a supporting electrolyte (Michael et al., 2015). The shapes and sizes of the obtained morphological features depended on the ABSA reactivity and AN concentration. Atomic force microscopy (AFM) data indicated the formation of globular micro/nanostructures for all the examined copolymers, P(An-co-ABSA) that exhibit also good protective performance against the corrosion of SS.

6.3.2 MULTILAYER CP COATINGS

Multilayer coatings of CPs have been used for the modification of the metal or alloy surface to ensure a feasible deposition of other polymers acting as protective coatings. This approach has been used in the case of AN electropolymerization on Zn and mild steel from aqueous media (Hasanov & Bilgic, 2009; Iroh et al., 2003; Lacroix et al., 2000). A thin PPy film was first deposited on the metal substrate from a neutral salicylate medium. The PPy thin layer behaves as a noble metal layer slowing down the substrate corrosion rate. Then, PAN was electrodeposited on the metal–PPy substrate from an acidic medium. The PPy/PAN bilayers and composites of PPy-PAN electrodeposited in one-step process (Pruna & Pilan, 2012; Rajagopalan & Iroh, 2001) exhibit superior anticorrosion performance in sodium chloride solutions than the plain PAN or PPy layers do. A finding that points to the mechanism by which the protection efficiency of the PPy/PAN bilayer is greatly improved is that when PAN was first deposited on steel, instead of PPy, the resulting PAN/PPy has a lesser protective ability than the PPy/PAN bilayer does (Hasanov & Bilgic, 2009). It seems that the PPy film provides a better barrier effect and sufficient adherence to the top PAN layer. The strip-shaped PAN formed within the matrix of PPy lowered the permeability of the PPy/PAN coating and, therefore, the transport of electrolyte, O_2 and H_2O, through the coating. PPy and PAN bilayers electrochemically synthesized on carbon steel and SS of type AISI 304 exhibit superior protective performance than the corresponding homopolymers (Panah & Danaee, 2010; Tan & Blackwood, 2003).

Bilayers of PAN and poly(N-methylaniline) (PNMA) were synthesized electrochemically on mild steel surface in a layer-by-layer manner from oxalic acid solutions (Narayanasamy & Rajendran 2010) as well as of PPy and PNMA from oxalic acid solutions containing DoS (Zeybek et al. 2011). In the former case, a galvanic interaction of CP bilayers was considered as the origin of the better corrosion resistance observed in 3.5 wt% NaCl solution in comparison with the corresponding homopolymer coatings. The metal/PAN/PNMA shows better stability than the metal/PNMA/PAN. In the latter case, the PNMA/PPy-DoS coating exhibited the best anticorrosion performance in highly aggressive 0.5 M HCl solution in comparison with the corresponding homopolymer coatings and PPy-DoS/PNMA coating.

The corrosion protection mechanism of these coatings involves the formation of a stable interphase between the coating and the metal substrate during electrodeposition, as well as the very low anionic permeability of the coating due to the doping of PPy by DoS. Relatively high protective efficiency was also observed for the homopolymer PPy-DoS coating. These examples show the importance of diminishing the anionic permeability of the coating, as the presence of the dodecylsulfate ion was found to be the cause of improved corrosion resistance performance.

In another attempt, multilayer thin films of poly(vinylsulfonic acid, sodium salt) (PVSS) and PAN were synthesized using a layer-by-layer technique on the surface of AA 2024 (Gomes & Oliveira, 2011). Several layers of PVSS/PAN were synthesized and were tested against corrosion. It was found that optimum anticorrosion performance was obtained at eight layers of PVSS/PAN on the surface of AA 2024 in the chloride media.

It should be noted that metal/CP bilayers may also offer improved corrosion protection. PPy electropolymerized on the surface of a copper layer deposited on aluminum imparted greater corrosion resistance than just copper-deposited aluminum and PPy-deposited aluminum did (Mert et al., 2011). This improvement was found to be a result of the enhanced barrier protection from the presence of various multilayers of CPs and metal.

6.3.3 PAN Oligomers

PAN exhibits greater diversity in its electrochemistry and is extensively used in corrosion protection. As was mentioned previously, its high conductivity, readily achieved by protonation, along with its easy preparation, makes it a more promising material among CPs in coating technology. On the other hand, the low solubility and processability of PAN limits its use in protective coatings, as in many other applications. Among several strategies suggested to improve the processability of PAN is the synthesis of AN oligomers (AOs). Thus, during the last decade, there is a growing interest for lower molecular weight CPs that have better solubility, processability, enhanced thermal stability whereas they retain their conductivity to high levels, similar to those of a high molecular weight polymer. AOs have currently attracted a special scientific interest in corrosion protection.

The synthesis of AOs with well-defined structure and end-groups has been reported. AOs were synthesized by a compensation polymerization from macromonomers of An oligomers such as polyimide (Huang et al., 2009a) and epoxy resin (Huang et al., 2009b). The oxidative coupling of AN oligomers and 1,4-phenylenediamine (Chao et al., 2007) was shown to be a relatively simple way (Figure 6.4) to synthesize various AOs-based electroactive coatings on cold-rolled steel (CRS) with enhanced anticorrosion ability.

During the last decade, there has been intense interest in lower-molecular-weight polymers that have better solubility, processability, and enhanced thermal stability, as they retain their conductivity to high levels, similar to those of a high-molecular-weight polymer. Aniline oligomers (AOs) are currently attracting special scientific interest in corrosion protection. By increasing the number of AN units from three to eight, the conductivity increases with the intact AN unit such that the heptamer

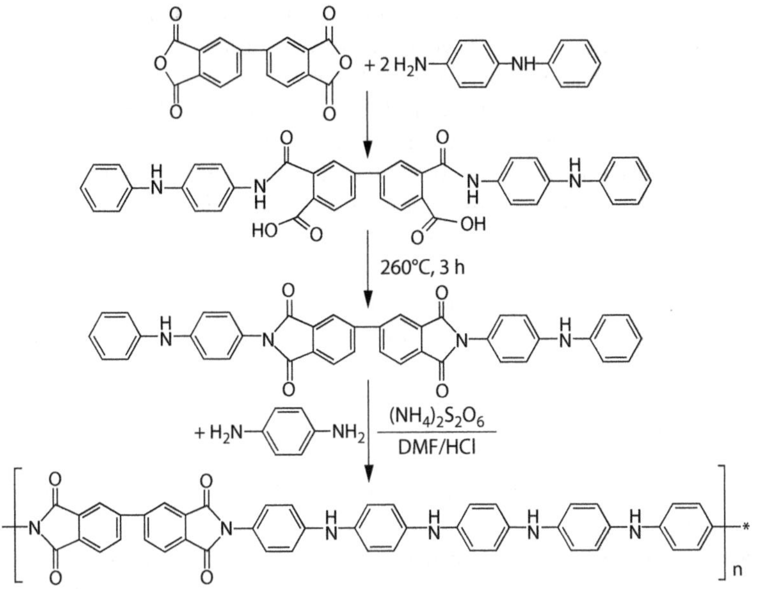

FIGURE 6.4 Schematic representation of the synthesis of An trimers and polyimides. (From Chao, D., Cui, L., Lu, X., Mao, H. S., Zhang, W. M., Wei, Y., *Eur. Polym. J.*, *43*, 2641–2647, 2007.)

exhibits conductivity approaching that of polymeric PAN. Recent experimental evidence shows that protection of the metallic substrate provided by AO is connected mainly with the change in the oxygen-reduction kinetics from a four- to a two-electron path on the AO-covered mild steel. As a consequence, the formed hydrogen peroxide may oxidize during corrosion Fe^{2+} to Fe^{3+}, resulting in the formation of $Fe(OH)_3$ in the AO pores (Grgur, 2014). The formation of $Fe(OH)_3$ reduces the corrosion rate and may explain the self-healing ability of metal coatings consisting of AO and perhaps of thin CP films. Thin PAN films often formed electrochemically are considered to be more likely AO than real PAN (Gospodinova & Terlemezyan, 1998). Several researchers have shown the protective efficiency of AO in chloride-containing solutions either deposited alone on low-carbon steel (Yadav et al., 2013) or incorporated into epoxy thermosets (Huang et al., 2009a), polyamide (Huang et al., 2012), polyimide (Huang et al., 2009a, 2011a,b), and polyurethane (Huang et al., 2013) deposited on CRS. The AO deposited on iron surfaces modifies the AO–iron oxide interface, leading to a band bending in the oxide and a decrease in the substrate work function (Greiner et al., 2008).

The low solubility and processability of PAN limit its applications in many areas and anticorrosion technology. Therefore, several strategies have been suggested to improve the processability of PAN. Among them, the synthesis of An oligomers (AOs) with well-defined structure and end-groups has been reported. AOs were synthesized by a compensation polymerization from macro-monomers of An oligomers such as polyimide (Huang et al., 2009a) and epoxy resin (Huang et al., 2009b). The oxidative coupling of AN oligomers and 1,4-phenylenediamine (Chao et al., 2007) was shown to be a relatively simple way (Figure 6.4) to synthesize various AO-based electroactive coatings with enhanced anticorrosion ability.

AOs-based electroactive coatings with different AN units were investigated with respect to their protective ability against corrosion of cold-rolled steel in chloride-containing solutions. For instance, the protective performance of AN pentamer-based electroactive polyimide (AP-based EPI) coating on CRS evaluated in 3.5 wt% NaCl solution was found to be superior compared with that of the non-electroactive (NEPI) counterpart coating (Huang et al., 2011b).

Huang et al. (2009a, 2011a,b) prepared AO-based electroactive coatings with different AN units and investigated their protective ability against corrosion of CRS in chloride-containing solutions. For instance, the protective performance of An pentamer-based electroactive polyimide (AP-based EPI) coating on CRS evaluated in 3.5 wt% NaCl solution was found to be superior compared with that of the nonelectroactive counterpart coating (Huang et al., 2011b). Although an enhanced physical barrier effect is ensured by the AP-based EPI coatings applied on CRS, the improved protection properties can be attributed to the redox catalytic properties of the AN pentamer in the AP-based EPI coating through an anodic mechanism. AOs induce stable passivation of the metal substrate as evidenced by scanning electron microscopy (SEM) (Figure 6.5a,b) and electron spectroscopy for chemical analysis (ESCA) investigations. ESCA plots of CRS after its immersion in the 3.5 wt% NaCl solution show that the passive oxide layer is composed of an outer layer of Fe_2O_3 and an inner layer of Fe_3O_4. The anodic protection mechanism, shown schematically in Figure 6.5c, seems to explain sufficiently the improved anticorrosion performance of EPI in neutral media.

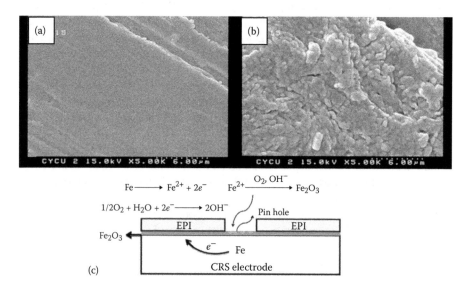

FIGURE 6.5   (See color insert.) SEM images of the (a) as-polished bare CRS and (b) the EPI-coated CRS. (c) Schematic presentation of the mechanism by which the AP-based EPI-coated CRS is protected against corrosion when immersed in 3.5 wt% NaCl. (From Huang, T.-C., Yeh, T.-C., Huang, T.-C., Ji, W.-F., Chou, Y.-C., Hung, W.-I., Yeh, J.-M., Tsai, M.-H., *Electrochim. Acta, 56*, 10151–10158, 2011.)

The oligomeric AN-based EPI (AN-capped AN trimer) combined with organophilic clay platelets leads to a polyiimde-clay nanocomposite (PCN) coating, which indicates good anticorrosion performance for the CRS in 5 wt% NaCl solution (Huang et al., 2011a). This PCN coating can be classified as an advanced anticorrosion material, in which synergistic effects, namely, the redox catalytic activity of AN trimer along with the ability of well-dispersed clay platelets, increase effectively the length of diffusion pathways for oxygen and water and decrease the permeability of the coating.

Furthermore, electroactive waterborne polyurethane (EWPU) containing conjugated segments of electroactive aminocapped aniline trimer (ACAT) unit was also applied on CRS and investigated with respect to corrosion protective properties in 3.5 wt% NaCl solution (Huang et al., 2013). The EWPU-based coating can be classified as an advanced eco-friendly coating with enhanced thermal stability that functions as protection coating via an anodic protection mechanism inducing the formation of stable passive oxide layer as indicated by SEM and X-ray photoelectron spectroscopy (XPS) studies.

6.4 HYDROPHOBIC AND SUPERHYDROPHOBIC CP COATINGS

The penetration of electrolyte into CP-based coatings is an essential cause of their failure in efficiently protecting metals against corrosion. Electrolyte and water penetration is dependent on the surface wetting ability. Surface wettability is highly dependent on the intrinsic hydrophobicity of coatings and their roughness geometry. CPs have unique properties allowing tuning of their surface wettability, for example, by reversibly incorporating various hydrophobic/hydrophilic doping ions, by modifying the molecular CP structure through functionalization with various hydrophobic/hydrophilic substituents, by changing the nature of the polymerizable core by combining CPs with other organic/inorganic materials (hybrid coatings or composites), or by utilizing different polymerization techniques (Darmanin & Guittard, 2013).

Surface wetting is the ability of a liquid to maintain contact with a solid surface depending on the intermolecular interactions between the liquid and solid. The surface wettability (degree of wetting) depends on the surface energy and morphology/topology determined by a force balance between adhesive and cohesive forces. Adhesive forces between a liquid and a solid lead to a drop to spread across the surface, whereas cohesive forces cause the opposite effect, leading a drop to ball up and avoid contact with the surface. A measure of surface wettability is the apparent contact angle (θ) on a smooth, homogeneous, rigid, insoluble, and nonreactive surface given by the Young–Dupre equation,

$$\cos\theta^{\gamma} = (\gamma_{SV} - \gamma_{SL})/\gamma_{LV} \qquad (6.2)$$

where γ_{SV}, γ_{SL}, and γ_{LV} are the solid–vapor, solid–liquid, and liquid–vapor surface tensions as defined in Figure 6.6. In the case of water, the water contact, should be $<90°$ for an intrinsically hydrophilic surface, and $>90°$ for an intrinsically hydrophobic surface. Superhydrophilic surfaces are characterized with very small apparent contact angles ($<5°–10°$) and a rapid spreading of water, and inversely,

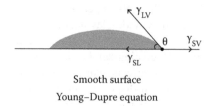

Smooth surface

Young–Dupre equation

FIGURE 6.6 Schematic representation of a water droplet on a smooth surface, following the Young–Dupre equation. (From Darmanin, T. Guittard, F., *Prog. Polym. Sci.*, *38*, 656–682, 2013.)

superhydrophobic surfaces are characterized by apparent contact angle higher than 150°. Moreover, the hysteresis, H, and sliding angle, α, both provide information on the adhesion between the droplet and the surface. The hysteresis is defined as the difference between the advancing (θ_a) and receding (θ_r) contact angles, $H = \theta_a - \theta_r$. It is a measure of the stickiness of the surface. The advancing and receding contact angles are taken just before the droplet rolls off the surface. When the surface inclination exceeds a certain value α, a droplet placed on the surface can roll off it.

The Wenzel and Cassie–Baxter equations (Darmanin & Guittard, 2013; Darmanin et al., 2013; Marmur, 2013), which are derived from the Young–Dupre equation, describe how the increase or decrease in surface hydrophobicity is possible by creating a surface roughness or porosity. The Wenzel equation assumes that when a water droplet is deposited onto a rough surface, it completely wets the underlying surface, as seen in Figure 6.7a. A roughness parameter is introduced in the Young–Dupre equation, leading to the Wenzel equation:

$$\cos\theta = r\cos\theta^Y \qquad (6.3)$$

The Wenzel equation can predict both the possibilities of reaching superhydrophilic (Figure 6.7a) and superhydrophobic (Figure 6.7b) properties. Indeed, Equation 6.3 shows that, if the surface is intrinsically hydrophobic ($\theta_w^Y < 90°$), increasing the surface roughness increases the surface hydrophobicity, and reversely, if the surface is intrinsically hydrophilic ($\theta_w^Y < 90°$), decreasing the surface roughness decreases the surface hydrophobicity.

Therefore, the wettability of surfaces is governed by three main parameters: the liquid-repellent properties of the compounds present at the outer surface, the surface roughness, and the surface morphology/topography. It is worth mentioning that the surface morphology and the surface roughness are clearly two different parameters. The surface morphology represents the shape of the structures present on the surfaces, such as pillar, fibers, cauliflower-like structures, flower-like structures, or mushroom-like structures. This parameter is very important because it can induce different pinning effects and allows the trapping of various amounts of air, leading to various liquid-repellent properties (Figure 6.7). Hence, several morphologies can lead to the same roughness parameter but not to the same liquid-repellent properties.

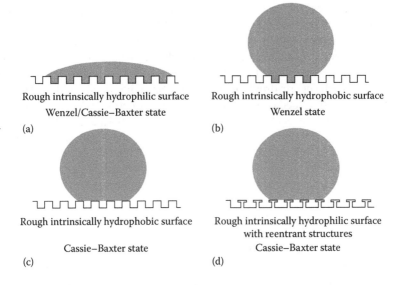

Rough intrinsically hydrophilic surface
Wenzel/Cassie–Baxter state

(a)

Rough intrinsically hydrophobic surface
Wenzel state

(b)

Rough intrinsically hydrophobic surface

Cassie–Baxter state

(c)

Rough intrinsically hydrophilic surface
with reentrant structures
Cassie–Baxter state

(d)

FIGURE 6.7 Schematic representation of a water droplet on (a) rough intrinsically hydro-philic surface (Wenzel/Cassie–Baxter state), (b) a rough hydrophobic surface (Wenzel state), (c) a rough intrinsically hydrophobic surface (Cassie–Baxter state), and (d) rough intrinsically hydrophilic surface with reentrant structures (Cassie–Baxter state). (From Darmanin, T., Guittard, F., *Prog. Polym. Sci.*, 38, 656–682, 2013.)

Roughness parameter and morphology can be combined in a unique parameter, as was introduced by Marmur (2013) in terms of the Cassie–Baxter theory.

The roughness parameter is important when the droplet is in the Wenzel state and the Wenzel equation can predict both the possibility of reaching superhydrophobic and superhydrophilic properties. However, the roughness parameter becomes less important in the Cassie–Baxter state (Darmanin & Guittard, 2013). Superhydrophobic properties can also be predicted by the Cassie–Baxter theory (Figure 6.6c). In this case, the water droplet is suspended on the top of the asperities and the air fraction present between the surface and the water droplet makes its suspension much easier. The interface is composed of the solid–liquid and liquid–vapor interfaces, and while the hydrophobicity increases with the air fraction, the hysteresis (H) and the sliding angle (α) decrease. The Cassie–Baxter equation is

$$\cos\theta = r_f f\cos\theta^Y + (1 - f)\cos\theta_\chi, \tag{6.4}$$

where f and $(1 - f)$ represent the solid and air fractions, respectively, and r_f is the roughness ratio of the wet area, as described by Marmur (2013), while θ_χ is the contact angle of a liquid on air: $\theta_\chi = 180°$ or $\cos\theta_\chi = -1$. The decrease in surface adhesiveness is a result of the increase in the liquid–vapor interface and, consequently, the decrease in the solid–liquid interface. Such surfaces are called "self-cleaning." It is considered that water droplet is in a Cassie–Baxter state if the sliding angle $\alpha <$ 10°. The hysteresis and sliding angles are essential in superhydrophobic surfaces, as the surface roughness increases the solid–liquid interface. The increase in adhesiveness in the Wenzel state results in the so-called "sticky" surfaces.

A schematic representation of a water droplet on a rough intrinsically hydrophilic surface (a, d) and a rough intrinsically hydrophobic surface (b, c) are depicted in Figure 6.7. If the water droplet fills the porosities, $\cos\theta_x$ becomes 1, which makes it possible to reach superhydrophilic properties (Shirtcliffe et al., 2005). Hence, the filling of the porosities decreases the contact angle. Moreover, when a droplet is in a Cassie–Baxter state, the state of the droplet can change from a Cassie–Baxter state to a Wenzel state by applying an external pressure (Bormashenko, 2010). When the Cassie–Baxter is highly stable, the surface is called "robust." For example, it was demonstrated that the presence of both surface microstructures and nanostructures, also named multiscale roughness, often increases the robustness of superhydrophobic surfaces.

Many theoretical and experimental studies were devoted to investigate the possibilities of reaching superhydrophobic surfaces from hydrophilic materials ($\theta^{\gamma}_{water} < 90°$) and, as a consequence, superoleo-phobic surfaces from oleophilic materials ($\theta^{\gamma}_{oils} < 90°$) by using or modifying the Cassie–Baxter equation (Marmur, 2013). Models were established and a necessary condition was found to be the presence of specific roughness geometries named "reentrant" structures, including "overhanging" or "mushroom-like" structures, in which the air trapped (Figure 6.7d) may induce a negative Laplace pressure difference (Wu & Suzuki, 2011). Hence, an energy barrier is created between the Wenzel and the Cassie–Baxter. As a consequence, the water droplet is in a metastable Cassie–Baxter state and the application of an external pressure can induce the switching between the two states.

To sum up, two parameters are very important to control the surface wettability of a CP coating, namely, the intrinsic liquid-repellent properties of the materials and the roughness geometry. The next sections will be dedicated to the control of the surface wettability of CP coatings by controlling these parameters. Wettability is essentially reduced by doping of CPs by hydrophobic ions. However, extreme water repellency is observed only in rough hydrophobic surfaces displaying the property of superhydrophobicity ($\theta^{\gamma}_w \geq 150°$). Control of the morphology of the surface in the micron and nanometer scales is the key to achieve surfaces of such a low wettability.

6.4.1 DOPING WITH HYDROPHOBIC IONS

Surface wettability and, hence, penetration of water and/or electrolyte can be remarkably reduced by incorporating hydrophobic ionic groups into CPs that enable the coating surface to repel water. Because of the ease of CP polymerization and control of the doping–dedoping processes upon imposing suitable potential/current values, reversible switching between superhydrophobicity and supehydrophilicity of CPs seems promising to be exploited for optimizing approaches used in anticorrosion technology.

Doping of CPs with hydrophobic ions can lead to superhydrophobic surfaces if the surfaces are sufficiently structured. These doping ions can be easily removed by switching the surface wettability from superhydrophobic to superhydrophilic.

Because the "natural" form of CPs is the dedoped (reduced) state, a main drawback of this approach is that the surfaces release the doping anions after aging, and hence, the stability of the doped state is highly dependent on the molecular structure of CP. For instance, the electrochemical growth of PAN in the presence of perfluorooctane-sulfonic acid led to submicron PAN fibers with helical structure (Xu et al., 2008). A conformation of the fibers is observed because of the presence of the perfluorinated chains, known to adopt helical structures, and the presence of the sulfonate groups, which increase the interaction by hydrogen bonding. Superhydrophobic properties with low hysteresis ($\theta_w^Y = 153°$, $H = 8°$) were reported for these surfaces with the possibility to switch from superhydrophobic to superhydrophilic by dedoping.

Perfluorooctanesulfonate anions were also used to obtain superhydrophobic micro/nanostructured PPy (Chang & Hunter, 2011; Liu et al., 2010; Mecerreyes et al., 2002; Xu et al., 2005) and poly(3,4-ethylenedioxythiophene) films with switchable wettability (Wolfs et al., 2011). For example, superhydrophobic PPy films containing a perfluoronated dopant, the tetraethylammonium perfluorooctanesulfonate ($Et_4N^+CF_3(CF_2)_7SO_3^-$, TEAPFOS) were synthesized by electrochemical (galvanostatic) polymerization, in the presence of Fe^{3+} ($<0^{-3}$ M) acting as a catalyst, on Au-coated glass from Py-containing acetonitrile solutions. These resulting PPy films exhibit an extended porous structure, in contrast to the compact films obtained in the absence of Fe^{3+}. By holding the perfluorooctanesulfonate (PFOS)-doped PPy films (oxidized state) in 0.05 TEAPFOS acetonitrile solution at a negative potential (-0.6 V vs. Ag/AgCl) for 20 min, the oxidized PPy was converted to neutral PPy. Neutral PPy films were then washed in acetonitrile and dried under a flow of argon. By holding the neutral PPy films in the solution at a positive potential (1.0 V vs. Ag/AgCl) for 20 min, the neutral PPy was converted to the oxidized PFOS-doped PPy. Oxidized PPy films were then washed in acetonitrile and dried under a flow of argon. The switching experiments can be conducted repeatedly on the same sample.

The profile of water contact angle for highly porous PFOS-doped PPy synthesized in the presence of Fe^{3+} in its PFOS-doped state (oxidized) shows superhydrophobicity with a water contact angle of $152° \pm 2°$. The neutral PPy film in its undoped state (reduced) shows also superhydrophilicity with a water contact angle of $\sim 0°$. The corresponding profile for the oxidized and reduced states of the compact PPy films synthesized without Fe^{3+} was also examined comparatively; the oxidized PPy film shows hydrophobicity with a water contact angle of $105° \pm 2°$, while the neutral PPy film seems hydrophilic with a water contact angle of $48° \pm 2°$ (Xu et al., 2005). This was attributed to the extended porous structure (pore size ranges from 10 to 50 µm) of the PFOS-doped PPy synthesized in the presence of Fe^{3+} (Figure 6.8a,b). In the absence of Fe^{3+} (Figure 6.8a,b) the pores were smaller in size, in the range of 1–4 µm, surrounded by submicron particles with diameters of 0.5–1 µm. This type of structure, with roughness in both coarse and fine scales, is more effective in achieving superhydrophobicity and superhydrophilicity.

Coating of active metals with such porous PFOS-doped PPy might improve metal corrosion resistance in comparison with the hydrophobic compact PPy films. The corrosion protection of a coating depends, among others, on (1) water adsorption by the coating, (2) transport of water in the coating, and (3) the accessibility of water to the coating/metal substrate interface. It is therefore plausible to assume that the

FIGURE 6.8 SEM images of PPy films synthesized in the presence (a, b) and absence of Fe^{3+} (d, e). (From Xu, L., Chen, W., Mulchandani, A., Yan, Y., *Angew Chem. Int. Ed.*, *44*, 6009–6012, 2005.)

superhydrophobic CPs with low wettability would be able to effectively prevent the water from absorbing onto the substrate surface and therefore exhibit excellent corrosion resistance in wet environments.

Hydrophobic PPy films (water contact angle up to 125°) prepared on Zn substrates from aqueous solutions by galvanostatic deposition (200 mA/cm²), leading to 2 μm thickness for 3.5 sec, exhibit improved corrosion protection of Zn when 3,5-diisopropylsalylate was used as dopant instead of salicylate (Hermelin et al., 2008).

6.4.2 Superhydrophobic CP Structures
via a Morphology/Topology Control

Since 2000, several strategies have been developed to mimic nature and to improve hydrophobic effect and produce superhydrophobic surfaces from several substrates. This is usually achieved by an appropriate combination of the chemistry and morphology of a surface. Such a strategy involves the use of low surface energy materials such as fluorocarbon, hydrocarbon, and silicone materials. An approach to fabricate improved hydrophobic CP-based surfaces was shown via (1) curing the epoxy resin by amine-capped aniline trimer (ACAT) at room temperature in the absence of any solvent and (2) replicating multiscale papilla-like structures, observed on the surface of the *Xanthosoma sagittifolium* (*X. sagittifolium*) leaf, on the surface of the electroactive epoxy (EE) coating using polydimethylsiloxane (PDMS) as a negative template (Weng et al., 2012; Yang et al., 2012). The resulting hydrophobic EE (HEE) coating with the replicated nanostructured surface showed a hydrophobic characteristic with a water contact angle close to 120°. Increasing the amount of ACAT in the resulting EE accelerates the curing process, promotes thermal stability, and improves anticorrosion performance. The developed HEE coating exhibited superior anticorrosion performance in electrochemical corrosion tests as its corrosion rate diminishes by a factor of 450 as compared with that of the bare steel substrate (Yang et al., 2012). The improved corrosion protection is attributed to the synergistic effect of electroactivity and hydrophobicity from the HEE coatings, with the multiscale structures mimicking the surface of the *X. sagittifolium* leaf.

The combination of microstructured and nanostructured CPs with a superhydrophobic function has become an interesting subject in materials science. Peng et al. (2013) suggested that PAN surface with biomimetic superhydrophobic structures can be prepared by the nanocasting technique and applied in corrosion protection coatings, as can be seen schematically in Figure 6.9. PAN was synthesized first by conventionally oxidative polymerization of AN with ammonium persulfate (APS) as oxidant in 1.0 M HCl aqueous solution. Subsequently, PAN with superhydrophobic surface (SH-PAN) of biomimetic natural leaf was fabricated, using PDMS as negative template, through nanocasting onto CRS electrode.

The PDMS template prepared has negative *X. sagittifolium* leaf surface structures and is obtained after peeling the leaf off. Second, the substrate is covered with the PAN solution, and the template is pressed against the CRS. After the heating process and peeling off the PDMS template, an *X. sagittifolium* leaf-like surface is formed on the CRS. Figure 6.10 shows a photograph of natural fresh *X. sagittifolium* leaves and its SEM image in comparison with the PDMS negative template and *X. sagittifolium* leaf-like SH-PAN surface.

It was shown that the SH-PAN coating offered excellent water repellent properties because of the rough morphology on the surface (Figure 6.10), which allows trapping of various amounts of air within the valleys between hills. Therefore, the water or the aggressive Cl⁻ ions are prevented from reaching the CRS, and hence, the PAN

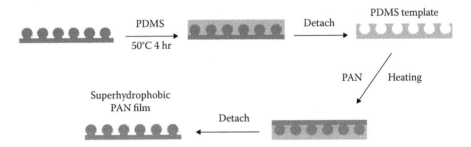

FIGURE 6.9 (See color insert.) Scheme for preparing the biomimetic superhydrophobic PAN film by using the nanocasting technique. (From Peng, C.-W., Chang, K.-C., Wenga, C.-J., Lai, M.-C., Hsua, C.-H., Hsua, S.-C., Hsua, Y.-Y., Hung, W.-I., Weid, Y., Yeh, J.-M., *Electrochim. Acta*, 95, 192–199, 2013.)

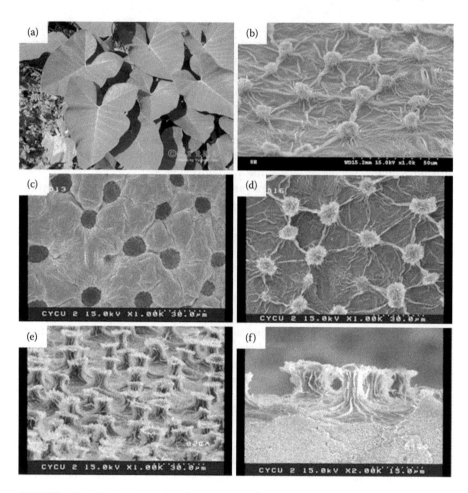

FIGURE 6.10 (See color insert.) (a) Photograph of the *X. sagittifolium* leaves, (b) SEM image of the natural leaf, (c) the PDMS negative template, (d) *X. sagittifolium* leaf-like SH-PAN surface, (e) a section view of the polymeric surface, and (f) magnification cross-section of SH-PAN. (From Peng, C.-W, Chang, K.-C., Wenga, C.-J., Lai, M.-C., Hsua, C.-H., Hsua, S.-C., Hsua, Y.-Y., Hung, W.-I., Weid, Y., Yeh, J.-M., *Electrochim. Acta*, 95, 192–199, 2013.)

reduction is avoided. SH-PAN-coated CRC displays improved corrosion resistance than bare CRS and flat PAN-coated CRS electrodes (Peng et al., 2013).

The use of electrodeposited superhydrophobic conducting PTh coating was found to effectively protect the underlying steel substrate from corrosion attack by first preventing water from being absorbed onto the coating, thus preventing the corrosive chemicals and corrosion products from diffusing through the coating, and second by causing an anodic shift in the corrosion potential as it galvanically couples to the metal substrate. Standard electrochemical measurements revealed that the steel coated with antiwetting nanostructured PTh film is very well protected from corrosion for up to 7 days when it was immersed in chloride solution of different pH and temperature. Its protection efficiency was greater than 95%. Fabrication of the dual properties of superhydrophobic anticorrosion nanostructured CP coating follows a two-step coating procedure that is very simple and can be used to coat any metallic surface (de Leon et al., 2012).

6.4.3 SUPERHYDROPHOBIC CP-BASED COMPOSITES/NANOCOMPOSITES

Advanced anticorrosion coating materials with synergistic effect of superhydrophobicity properties (θ_w^Y ca ~161°) and redox catalytic capability were prepared from fluoro-substituted PAN (F-PAN) incorporated with methyl triethoxysilane (MTMS)-base silsesquioxane spheres (Weng et al., 2012). F-PAN was first synthesized by conventionally oxidative polymerization with APS as oxidant. On the other hand, silsesquioxane spheres with a diameter of ~300 nm were prepared by the (Stöber 1968) with MTMS as a precursor (Sankaraiah et al., 2008). Subsequently, the silsesquioxane spheres and F-PAN were blended into NMP, followed by casting onto CRS substrate to give superhydrophobic surface (SHS) coating material a thickness of ~6 μm. The protective performance of the resulting SHS coating is compared hydrophobic surface (HS) coating material in saline solutions. Based on a series of electrochemical measurements, it was shown that the as-prepared coating materials exhibited superior corrosion protection effect on CRS electrode and provided an effective barrier to aggressive species. Compared with the protection efficiency between F-PAN coating and SHS coating, the SHS showed a protection efficiency (P_{EF}) value of 95.92%, which was greater than that of smooth F-PAN coating (P_{EF} = 79.76%) (Weng et al., 2012).

Enhancement of the corrosion resistance of the SHS-coated CRS electrode as compared with that of HS-coated CRC can be interpreted by considering that the synergistic effect of the water repellency offered a water barrier in the coating surface and the electroactivity of the PAN provided anodic protection to the CRS substrate in the corrosive medium. However, since F-PAN is present in both HS- and SHS-coated CRC surfaces, a major reason for the enhancement of the corrosion protection of the SHS-coated CRC could be the water repellent property of the SHS coating, which prevents effectively aggressive species transported by water from reaching the substrate.

The surface wettability and morphology of bare, F-PAN-, and SHS-coated CRS samples were characterized using water-contact angle meter, SEM, and atomic force microscopy (AFM). Figure 6.11a shows the surface morphology of bare CRS substrate and the image of the corresponding water contact angle. It reveals that the surface of bare CRS substrate is considerably smooth with a water θ_w^Y ~74.1°. Figure 6.11b illustrates the surface morphology of CRS substrate treated with smooth, low-surface-energy, F-PAN coating with an increased water contact angle (θ_w^Y ~95°) compared with that of the bare CRS substrate. A significant increase in the water contact angle (θ_w^Y ~161°) is observed when the CRC substrate was coated with F-PAN incorporated with silsesquioxane particles indicating a superhydrophobic surface (Figure 6.11c). Indeed, the SEM image of Figure 6.11c shows a rougher surface for the SHS-coated CRS samples since the silsesquioxane particles have assembled together due to the binding induced by F-PAN. The morphology of the F-PAN was smooth as can be seen in the 3-D AFM image of Figure 6.11d. The surface roughness Ra (arithmetic

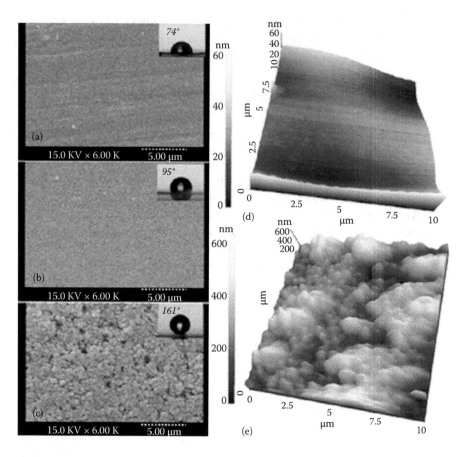

FIGURE 6.11 **(See color insert.)** Correlation of surface roughness and water contact angle on bare and coated CRS surfaces. (a) SEM image of bare-CRS, (b) SEM image of F-PAN on CRS, (c) SEM image of SHS on CRS, (d) AFM image of F-PAN on CRS, and (e) AFM image of SHS on CRS. In the top-right corner of the SEM images, optical images of the corresponding water contact angles are depicted. (From Weng, C.-J., Chang, C.-H., Lin, I. L., Yeh, J.-M., Wei, Y., Hsu, C.-L., Chen, P.-H., *Surf Coat Technol.*, *207*, 42–49, 2012.)

average) was found to be Ra~7.90 nm. However, as Figure 6.11e shows, the surface becomes quite rough (Ra~90.25 nm) with a hill-like structure when the CRS was covered with the SHS coating (F-PAN with silsesquioxane particles). These results indicate that the surface wettability of the solid surfaces is governed not only by the surface energy of the material but also by the surface roughness of solid.

Enhancement of the hydrophobicity of CP-based composites of micro- and nanoscale can be achieved by using different inorganic particles and CP. In general, the hydrophobic PAN–SiO$_2$ nanocomposites (HPSC) synthesized chemically and deposited on mild steel exhibit improved protective efficiency as compared with PAN (Bhandari et al., 2012). The HPSC coating acts as a barrier between the metal and the corrosive environment, preventing the chloride ion penetration into the coating owing to the increased hydrophobic character of the composite. On the other hand, HPSC results in the formation of a passive oxide film on mild steel. It is also admitted that the presence of SiO_2 nanoparticles leads to the reinforcement of PAN and, thereby, to the diminution of its degradation.

Although the role of CP superhydrophobicity on metal corrosion protection is unclear and yet under investigation, structured coatings based on replicating fresh plant leaves (Weng et al., 2011; Peng et al., 2012; Yang et al., 2012) or oxides (Xu et al., 2011; Boinovich et al., 2012) support the view that the coating superhydrophobic effect may improve the CRS protection in corrosive media. However, a common phenomenon reported in the literature is that superhydrophobicity can be maintained for a long time only in air, while it turns to hydrophobicity within a few hours after immersion in aqueous corrosion media, indicating poor stability and short duration of superhydrophobicity. A question posed by Yu and Tian (2014) is whether superhydrophobicity is indeed better than hydrophobicity for anticorrosion coatings. Evaluation results, reported for several composite coatings with superhydrophobic and hydrophobic surfaces, have shown that superhydrophobic surfaces do not always provide better anticorrosion protection. It was suggested that this behavior is linked to subchannels in the composite coating. Such subchannels in the superhydrophobic surface are more than those in the hydrophobic surface due to the difference of surface roughness. Prepared samples of carbon steel coated with a multilayered composite were tested comparatively via a Lotus experiment (with a continuous seawater dropping up to 72 h) and an immersion experiment (immersion in seawater for 72 h). Electrochemical results have shown that the surface of the immersed sample has changed from superhydrophobic to hydrophobic as θ_w^Y decreased from 156° to 130°, while the Lotus sample maintains superhydrophobicity even after 72 h seawater dropping (θ_w^Y >150°), which demonstrated that the superhydrophobic sample is effective if its rolling characteristics are maintained (Yu & Tian, 2014). Therefore, superhydrophobicity does not definitely mean better anticorrosion performance than hydrophobicity. The combination of micro- and nano-structured CP and particles of metals, oxides, organic, and biological materials to prepare composites with controllable morphologies, sizes, and hydrophobic/superhydrophobic properties is still challenging. It can be expected that such efficient synthetic paths may lead to notable improvements in the anticorrosion performance of CP-based composites and nanocomposites.

REFERENCES

Adenier, A., Cabet-Deliry, E., Lalot, T., Pinson, J., & Podvorica, F. (2002). Attachment of polymers to organic moieties covalently bonded to iron surfaces. *Chem. Mater.*, *14*, 4576–4586.

Allonque, P., Delamar, M., Desbat, B., Fagebaume, O., Hitmi, R., Pinson, J., & Saveant, J. M. (1997). Covalent modification of carbon surfaces by aryl radicals generated from the electrochemical reduction of diazonium salts. *J. Am. Chem. Soc.*, *119*, 201–207.

Barisci, J. N., Lewis, T. W., Spinks, G. M., Too, C. O., & Wallace, G. G. (1998). Conducting polymers as a basis for responsive materials systems. *J. Intell. Mater. Syst. Struct.*, *9*, 723–731.

Bhandari, H., Srivastav, R., Choudhary, V., & Dhawan, S. K. (2010). Enhancement of corrosion protection efficiency of iron by poly(aniline-co-amino-naphthol-sulphonic acid) nanowires coating in highly acidic medium. *Thin Solid Films*, *519*(3), 1031–1039.

Bhandari, H., Choudhary, V., & Dhawan, S. K. (2011). Influence of self-doped poly(aniline-co-4-amino-3-hydroxy-naphthalene-1-sulfonic acid) on corrosion inhibition behaviour of iron in acidic medium. *Synth. Met.*, *161*(9–10), 753–762.

Bhandari, H., Anoop Kumar, S., & Dhawan, S. K. (2012). Conducting polymer nanocomposites for anticorrosive and antistatic applications. In F. Ebrahimi (Ed.), *Nanocomposites—New Trends and Developments*. Rejeka, Croatia:INTECHOPEN.COM: INTECH.

Boinovich, L. B., Gnedenkov, S. V., Alpysbaeva, D. A., Egorkin, V. S., Emelyanenko, A. M., Sinebryukhov, S. L., & Zaretskaya, A. K. (2012). Corrosion resistance of composite coatings on low-carbon steel containing hydrophobic and superhydrophobic layers in combination with oxide sublayers. *Corros. Sci.*, *55*, 238–245.

Bormashenko, E. (2010). Wetting transitions on biomimetic surfaces. *Phil. Trans. R. Soc. A*, *368*, 4695–4711.

Çakmakcı, İ., Duran, B., Duran, M., & Bereket, G. (2013). Experimental and theoretical studies on protective properties of poly(pyrrole-co-N-methyl pyrrole) coatings on copper in chloride media. *Corros. Sci.*, *69*, 252–261.

Chang, J. H., & Hunter, W. (2011). A superhydrophobic to superhydrophilic in situ wettability switch of microstructured polypyrrole surfaces. *Macromol. Rapid Commun.*, *32*, 718–723.

Chao, D., Cui, L., Lu, X., Mao, H. S., Zhang, W. M., & Wei, Y. (2007). Electroactive polyimide with oligoaniline in the main chain via oxidative coupling polymerization. *Eur. Polym. J.*, *43*, 2641–2647.

Darmanin, T., & Guittard, F. (2013). Wettability of conducting polymers: From superhydrophilicity to superoleophobicity. *Prog. Polym. Sci.*, *38*, 656–682.

Darmanin, T., Taffin de Givenchy, E., Amigoni, S., & Guittard, F. (2013). Superhydrophobic surfaces by electrochemical processes. *Adv. Mat.*, *25*, 1378–1394.

de Leon, A. C. C., Pernites, R. B., & Advincula, R. C. (2012). Superhydrophobic colloidally textured polythiophene film as superior anticorrosion coating. *ACS Appl. Mater. Interfaces*, *4*, 3169–3176.

Elhalawany, N., Mossad, M. A., & Zahran, M. K. (2014). Novel water based coatings containing some conducting polymers nanoparticles (CPNs) as corrosion inhibitors. *Prog. Org. Coat.*, *77*(3), 725–732.

Ferreira, C. A., Aeiyach, S., Aaron, J. J., & Lacaze, P. C. (1996). Electrosynthesis of strongly adherent Ppy coatings on iron and mild steel in aqueous media. *Electrochim. Acta*, *41*, 1801–1809.

Gomes, E. C., & Oliveira, M. A. S. (2011). Corrosion protection by multilayer coating using layer-by-layer technique. *Surf. Coat. Technol.*, *205*(8–9), 2857–2864.

Gospodinova, N., & Terlemezyan, L. (1998). Conducting polymers prepared by oxidative polymerization: Polyaniline. *Prog. Polym. Sci.*, *23*, 1443–1484.

Greiner, M. T., Festin, M., & Kruse, P. (2008). Investigation of corrosion-inhibiting aniline oligomer thin films on iron using photoelectron spectroscopy. *J. Phys. Chem. C*, *112*, 18991–19004.

Grgur, B. N. (2014). On the role of aniline oligomers on the corrosion protection of mild steel. *Synth. Met.*, *187*, 57–60.

Hasanov, R., & Bilgic, S. (2009). Monolayer and bilayer conducting polymer coatings for corrosion protection of steel in 1 M H2SO4 solution. *Prog. Org. Coatings*, *64*, 435–445.

Hermelin, E., Petitjean, J., Lacroix, J. C., Chane-Ching, K. I., Tanguy, J., & Lacaze, P. C. (2008). Ultrafast electrosynthesis of high hydrophobic polypyrrole coatings on a zinc electrode: Applications to the protection against corrosion. *Chem. Mater.*, *20*, 4447–4456.

Huang, J., & Wan, M. (1999). Polyaniline doped with different sulfonic acids by in situ doping polymerization. *J. Polym. Sci.: Part A: Polym. Chem., 37*, 1277–1284.

Huang, K.-Y., Jhuo, Y.-S., Wu, P.-S., Lin, C.-H., Yu, Y.-H., & Yeh, J.-M. (2009a). Electrochemical studies for the electroactivity of amine-capped aniline trimer on the anticorrosion effect of as-prepared polyimide coatings. *Eur. Polym. J., 45*, 485–493.

Huang, K.-Y., Shiu, C.-L., Wu, P.-S., Wei, Y., Yeh, J.-M., & Li, W.-T. (2009b). Effect of amino-capped aniline trimer on corrosion protection and physical properties for electroactive epoxy thermosets. *Electrochim. Acta, 54*, 5400–5407.

Huang, H.-Y., Huang, T.-C., Yeh, T.-C., Tsai, C.-Y., Lai, C.-L., Tsai, M.-H., Yeh, J.-M., & Chou, Y.-C. (2011a). Advanced anticorrosive materials prepared from aniline-capped aniline trimer-based electroactive polyimide clay nanocomposites materials with synergistic effects of redox catalytic capability and gas barrier properties. *Polymer, 52*, 2391–2400.

Huang, T.-C., Yeh, T.-C., Huang, T.-C., Ji, W.-F., Chou, Y.-C., Hung, W.-I., Yeh, J.-M., & Tsai, M.-H. (2011b). Electrochemical studies on aniline-pentamer-based electroactive polyimide coating: Corrosion protection and electrochromic properties. *Electrochim. Acta, 56*, 10151–10158.

Huang, T.-C., Yeh, T.-C., Huang, H.-Y., Ji, W.-F., Lin, T.-C., Chen, C.-A., Yang, T.-I., & Yeh, J.-M. (2012). Electrochemical investigations of the anticorrosive and electrochromic properties of electroactive polyamide. *Electrochim. Acta, 63*, 185–191.

Huang, H.-Y., Huang, T.-C., Lin, J.-C., Chang, J.-H., Lee, Y.-T., & Yeh, J.-M. (2013). Advanced environmentally friendly coatings prepared from amine-capped aniline trimer-based waterborne electroactive polyurethane. *Mater. Chem. Phys., 137*(3), 772–780.

Iroh, J. O., Zhua, Y., Shah, K., Levine, K., Rajagopalan, R., Uyar, T., Donley, M., Mantz, R., Johnson, J., Voevodin, N. N., Balbyshev, V. N., & Khramov, A. N. (2003). Electrochemical synthesis: A novel technique for processing multi-functional coatings. *Prog. Org. Coatings, 47*, 365–375.

Jaehne, E., Oberoi, S., & Adler, H.-J. P. (2008). Ultra thin layers as new concepts for corrosion inhibition and adhesion promotion. *Prog. Org. Coat., 61*, 211–213.

Kamaraj, K., Karpakam, V., Sathiyanarayanan, S., & Venkatachari, G. (2010). Electrosysnthesis of poly(aniline-co-m-amino benzoic acid) for corrosion protection of steel. *Mater. Chem. Phys., 122*(1), 123–128.

Kendig, M., Hon, M., & Warren, L. (2003). 'Smart' corrosion inhibiting coatings. *Prog. Org. Coat., 47*, 183–189.

Khalkhali, R. A., Price, W. E., & Wallace, G. G. (2003). Quartz crystal microbalance studies of the effect of solution temperature on the ion-exchange properties of polypyrrole conducting electroactive polymers. *React. Funct. Polym., 56*(3), 141–146.

Kinlen, P. J., Menon, V., & Ding, Y. W. (1999). A mechanistic investigation of polyaniline corrosion protection using the scanning reference electrode technique. *J. Electrochem. Soc., 146*(10), 3690–3695.

Kinlen, P. J., Ding, Y., & Silverman, D. C. (2002). Corrosion protection of mild steel using sulfonic and phosphonic acid-doped polyanilines. *Corrosion, 58*(6), 490–497.

Koehler, S., Ueda, M., Efimov, I., & Bund, A. (2007). An EQCM study of the deposition and doping/dedoping behavior of polypyrrole from phosphoric acid solutions. *Electrochim. Acta, 52*(9), 3040–3046.

Kosseoglou, D., Kokkinofta, R., & Sazou, D. (2011). FTIR spectroscopic characterization of NafionA (R)-polyaniline composite films employed for the corrosion control of stainless steel. *J. Solid State Electrochem., 15*(11–12), 2619–2631.

Kowalski, D., Ueda, M., & Ohtsuka, T. (2007). Corrosion protection of steel by bi-layered polypyrrole doped with molybdophosphate and naphthalenedisulfonate anions. *Corros. Sci., 49*(3), 1635–1644.

Lacaze, P. C., Ghilane, J., Randriamahazaka, H., & Lacroix, J.-C. (2010). Electroactive conducting polymers for the protection of metals against corrosion: From micro- to nanostructured films. In: A. Eftekhari, ed., *Nanostructured Conductive Polymers*. Chichester, UK: John Wiley & Sons, Ltd.

Lacroix, J. C., Camalet, J. L., Aeiyach, S., Chane-Ching, K. I., Petitjean, E., Chauveau, E., & Lacaze, P. C. (2000). Aniline electropolymarization on mild steel and zinc in a two-step process. *J. Electroanal. Chem.*, *481*, 76–81.

Liu, M., Nie, F., Wei, Z., Song, Y., & Jiang, L. (2010). In situ electrochemical switching of wetting state of oil droplet on conducting polymer films. *Langmuir*, *26*, 3993–3997.

Maia, G., Torresi, R. M., Ticianelli, E. A., & Nart, F. C. (1996). Charge compensation dynamics in the redox processes of polypyrrole–modified electrodes. *J. Phys. Chem.*, *100*(39), 15910–15916.

Marmur, A. (2013). Superhydrophobic and superhygrophobic surfaces: From understanding non-wettability to design considerations. *Soft Mater.*, *9*, 7009–7904.

Mecerreyes, D., Alvaro, V., Cantero, I., Bengoetxea, M., Calvo, P., Grande, H., Rodriguez, J., & Pomposo, J. (2002). Low surface energy conducting polypyrrole doped with a fluorinated counterion. *Adv. Mater.*, *14*, 749–752.

Mert, B. D., & Yazici, B. (2011). The electrochemical synthesis of poly(pyrrole-co-o-anisidine) on 3102 aluminum alloy and its corrosion protection properties. *Mater. Chem. Phys.*, *125*(3), 370–376.

Mert, B. D., Solmaz, R., Kardas, G., & Yazici, B. (2011). Copper/polypyrrole multilayer coating for 7075 aluminum alloy protection. *Prog. Org. Coat.*, *72*(4), 748–754.

Michael, K., Prochaska, C., & Sazou, D. (2015). Electrodeposition of self-doped copolymers of aniline with aminobenzensulfonic acids on stainless steel. Morphological and electrochemical characterization. *J. Solid State Electrochem.*, doi: 10.1007/s10008-015-2898-4.

Nagaoka, T., Nakao, H., Suyama, T., Ogura, K., Oyama, M., & Okazaki, S. (1997). Electrochemical characterization of soluble conducting polymers as ion exchangers. *Anal. Chem.*, *69*, 1030–1037.

Najari, A., Lang, P., Lacaze, P. C., & Mauer, D. (2009). A new organofunctional methoxysilane bilayer system for promoting adhesion of epoxidized rubber to zinc. *Prog. Org. Coat.*, *64*, 392–404.

Narayanasamy, B., & Rajendran, S. (2010). Electropolymerized bilayer coatings of polyaniline and poly(*N*-methylaniline) on mild steel and their corrosion protection performance. *Prog. Org. Coat.*, *67*(3), 246–254.

Ohtsuka, T. (2012). Corrosion protection of steels by conducting polymers. *Int. J. Corros.*, 2012, Article ID 915090, 7 pp., doi:10.1155/2012/915090.

Ozyilmaz, A. T. (2006). The corrosion performance of polyaniline film modified on nickel plated copper in aqueous p-toluenesulfonic acid solution. *Surf. Coat. Technol.*, *200*(12–13), 3918–3925.

Paliwoda-Porebska, G., Stratmann, M., Rohwerder, M., Potje-Kamloth, P., Lu, Y., Pich, A. Z., & Adler, H.-J. (2005). On the development of polypyrrole coatings with self-healing properties for iron corrosion protection. *Corros. Sci.*, *47*, 3216–3233.

Paliwoda-Porebska, G., Rohwerder, M., Stratmann, M., Rammelt, U., Duc, L. M., & Plieth, W. (2006). Release mechanism of electrodeposited polypyrrole doped with corrosion inhibitor anions. *J. Solid State Electrochem.*, *10*, 730–736.

Panah, N. B., & Danaee, I. (2010). Study of the anticorrosive properties of polypyrrole/polyaniline bilayer via electrochemical techniques. *Prog. Org. Coat.*, *68*(3), 214–218.

Paul, R. K., & Pillai, C. K. S. (2001). Melt/Processable polyaniline with functionalized phosphate ester dopants and its thermoplastic blends. *J. Appl Polym Sci*, 80, 1354–1367.

Peng, C.-W., Chang, K.-C., Weng, C.-J., Lai, M.-C., Hsu, C.-H., Hsu, S.-C., Hsu, Y.-Y., Hung, W.-I., Wei, Y., & Yeh, J.-M. (2013). Nano-casting technique to prepare polyaniline surface with biomimetic superhydrophobic structures for anticorrosion application. *Electrochim. Acta*, *95*, 192–199.

Pruna, A., & Pilan, L. (2012). Electrochemical study on new polymer composite for zinc corrosion protection. *Compos. Part B, 43,* 3251–3257.

Qi, K., Qiu, Y., Chen, Z., & Guo, X. (2012). Corrosion of conductive polypyrrole: Effects of environmental factors, electrochemical stimulation, and doping anions. *Corros. Sci., 60,* 50–58.

Rajagopalan, R., & Iroh, J. O. (2001). Development of polyaniline–polypyrrole composite coatings on steel by aqueous electrochemical process. *Electrochim. Acta, 46,* 2443–2455.

Reut, J., Opik, A., & Idla, K. (1999). Corrosion behavior of polypyrrole coated mild steel. *Synth. Met., 102*(1–3), 1392–1393.

Rohwerder, M. (2009). Conducting polymers for corrosion protection: A review. *Int. J. Mater. Res., 100*(10), 1331–1342.

Sankaraiah, S., Lee, J. M., Kim, J. H., & Choi, S. W. (2008). Preparation and characterization of surface-functionalized polysilsesquioxane hard spheres in aqueous medium. *Macromolecules, 41,* 6195–6204.

Santos, L. M., Ghilane, J., Fave, C., Lacaze, P.-C., Randriamahazaka, H., Abrantes, L. M., & Lacroix, J.-C. (2008). Electrografting polyaniline on carbon through the electroreduction of diazonium salts and the electrochemical polymerization of aniline. *J. Phys. Chem. C, 112,* 16103–16109.

Sazou, D., & Kosseoglou, D. (2006). Corrosion inhibition by Nafion (R)-polyaniline composite films deposited on stainless steel in a two-step process. *Electrochim. Acta, 51*(12), 2503–2511.

Sazou, D., & Kourouzidou, M. (2009). Electrochemical synthesis and anticorrosive properties of Nafion®–poly(aniline-co-o-aminophenol) coatings on stainless steel. *Electrochim. Acta, 54*(9), 2425–2433.

Shirtcliffe, N. J., McHale, G., Newton, M. I., Perrya, C. C., & Roacha, P. (2005). Porous materials show superhydrophobic to superhydrophilic switching. *Chem. Commun.,* 3135–3137.

Souza, S. (2007). Smart coating based on polyaniline acrylic blend for corrosion protection of different metals. *Surf. Coat. Technol., 201,* 7574–7581.

Stöber, W., Fink, A., & Bohn, E. (1968). Controlled growth of monodisperse silica sphere in the micron size range. *J. Collods Int. Sci., 26,* 62–69.

Tan, C. K., & Blackwood, D. J. (2003). Corrosion protection by multilayered conducting polymer coatings. *Corros. Sci., 45*(3), 545–557.

Tuken, T., Tansug, G., Yazici, B., & Erbil, M. (2007). Poly(N-methyl pyrrole) and its copolymer with pyrrole for mild steel protection. *Surf. Coat. Technol., 202,* 146–154.

Wang, J., Torardi, C. C., & Duch, M. W. (2007a). Polyaniline-related ion-barrier anticorrosion coatings I. Ionic permeability of polyaniline, cationic, and bipolar films. *Synth. Met., 157,* 846–850.

Wang, J., Torardi, C. C., & Duch, M. W. (2007b). Polyaniline-related ion-barrier anticorrosion coatings, II. Protection behavior of polyaniline, cationic, and bipolar films. *Synth. Met., 157,* 851–858.

Weng, C.-J., Chang, C.-H., Peng, C.-W., Yeh, J.-M., Hsu, C.-L., & Wei, Y. (2011). Advanced anticorrosive coatings prepared from the mimicked Xanthosoma sagittifolium-leaf-like electroactive epoxy with synergistic effects of superhydrophobicity and redox catalytic capability. *ACs Chem. of Mater., 23,* 2075–2083.

Weng, C.-J., Chang, C.-H., Lin, I. L., Yeh, J.-M., Wei, Y., Hsu, C.-L., & Chen, P.-H. (2012). Advanced anticorrosion coating materials prepared from fluoro-polyaniline-silica composites with synergistic effect of superhydrophobicity and redox catalytic capability. *Surf. Coat. Technol., 207,* 42–49.

Wolfs, M., Darmanin, T., & Guittard, F. (2011). Versatile superhydrophobic surfaces from a bioinspired approach. *Macromolecules, 44,* 9286–9294.

Wu, T., & Suzuki, Y. (2011). Design, microfabrication and evaluation of robust high-performance superlyophobic surfaces. *Sens. Actuators B, 156,* 401–409.

Xing, C., Zhang, Z., Yu, L., Waterhouse, G. I. N., & Zhang, L. (2014). Anti-corrosion performance of nanostructured poly(aniline-co-metanilic acid) on carbon steel. *Prog. Org. Coat.*, *77*(2), 354–360.

Xu, L., Chen, W., Mulchandani, A., & Yan, Y. (2005). Reversible conversion of conducting polymer films from superhydrophobic to superhydrophilic. *Angew. Chem. Int. Ed.*, *44*, 6009–6012.

Xu, L., Chen, Z., Chen, W., Mulchandani, A., & Yan, Y. (2008). Electrochemical synthesis of perfluorinated ion doped conducting polyaniline films consisting of helical fibers and their reversible switching between superhydrophobicity and superhydrophilicity. *Macromol. Rapid Commun.*, *29*, 832–838.

Xu, W., Song, J., Sun, J., Lu, Y., & Yu, Z. (2011). Rapid fabrication of large-area, corrosion-resistant superhydrophobic Mg alloy surfaces. *ACS Appl. Mater. Interfaces*, *3*, 4404–4414.

Yadav, D. K., Chauhan, D. S., Ahamad, I., & Quaraishi, M. A. (2013). Electrochemical behavior of steel/acid interface: Adsorption and inhibition effect of oligomeric aniline. *RSC Adv.*, *3*, 632–646.

Yalcinkaya, S., Tuken, T., Yazici, B., & Erbil, M. (2008). Electrochemical synthesis and characterization of poly(pyrrole-co-o-toluidine). *Prog. Org. Coat.*, *63*, 424–433.

Yalcinkaya, S., Tuken, T., Yazici, B., & Erbil, M. (2010). Electrochemical synthesis and corrosion behaviour of poly (pyrrole-co-o-anisidine-co-o-toluidine). *Curr. Appl. Phys.*, *10*(3), 783–789.

Yang, T.-I., Peng, C.-W., Lin, Y. L., Weng, C.-J., Edgington, G., Mylonakis, A., Huang, T.-C., Hsu, C.-H., Yeh, J.-M., & Weie, Y. (2012). Synergistic effect of electroactivity and hydrophobicity on the anticorrosion property of room-temperature-cured epoxy coatings with multi-scale structures mimicking the surface of *Xanthosoma sagittifolium* leaf. *J. Mater. Chem.*, *22*, 15845–15852.

Yu, D., & Tian, J. (2014). Superhydrophobicity: Is it really better than hydrophobicity on anti-corrosion? *Colloids Surf. A: Physicochem. Eng. Asp.*, *445*, 75–78.

Yu, D., Tian, J., Dai, J., & Wang, X. (2013). Corrosion resistance of three-layer superhydrophobic composite coating on carbon steel in seawater. *Electrochim. Acta*, *97*, 409–419.

Zeybek, B., Pekmez, N. O., & Kilic, E. (2011). Electrochemical synthesis of bilayer coatings of poly(*N*-methylaniline) and polypyrrole on mild steel and their corrosion protection performances. *Electrochim. Acta*, *56*(25), 9277–9286.

7 Outlook

Conducting polymers (CPs) for protection against the corrosion of metals and alloys have been utilized in a variety of formulations to ensure good adherence to the metal substrate and long-term functionality considering environmental factors. Evaluation of the protective efficiency of different CP-based coatings in various configurations shows very promising results, and owing to restrictions on the use of heavy metals and chromating processes as environmentally nonfriendly and toxic for human health, an important research field has developed during the last decades. Numerous studies have been published dealing, at the beginning, mostly with the protection mechanisms and giving emphasis lately to new strategies, which aim at improving protection efficiency considering also technical and economical factors.

Recent innovations in CP-based coatings are based on the technology of "smart" or "intelligent" coatings, terms used lately to describe multifunctional coatings, which can provide more than one function to the metal substrate. In this sense CP-based coatings belong to smart coatings and constitute the cornerstone for the development of a new technology. In fact, most of these coating technologies are currently under development in academia and industry. In particular, scaling up of innovative CP-based coatings for possible industrial applications is still in its infancy.

A CP film cannot alone offer complete protection to a metal. The simplest requirement to this configuration is the presence of a topcoat (i.e., another polymer film) to eliminate diffusion of corrosion species through the CP film toward the metal substrate and/or prevent ion exchange processes. Without a topcoat, within a few weeks, CP films fail to protect metals and alloys under constant immersion conditions in corrosive median, especially in the presence of chlorides. There are two basic methods to deposit CP films on oxidizable metal surfaces: electrochemical and chemical by using different CP formulations as dispersions/solutions or blends, for instance, with epoxy and acrylic resins. The selection of the suitable method depends on the advantages/disadvantages encountered by the specific multi-interfacial system of interest, namely, the metal surface/CP coating/corrosive environment.

Blending of CPs with epoxy or acrylic polymers takes advantage of the combination between the good mechanical properties and processability of conventional polymers with the reversible redox properties of CPs.

Future trends in the technology of CP-based coatings might be expected as follows:

- Preparation of advanced CP-based composites/nanocomposites with desired protective properties.
- Functionalized CP-based coatings of unique chemistry and nanostructure will be further explored owing to innovations in synthesis of conducting polymers, curing methods and combined utilization of inhibiting agents with CPs.

- Development of new surface pretreatments, to improve strong adherence for long-term protection with environmentally friendly methods, will be a popular research topic of the next decade.
- More attention will be placed to optimizing "smart" CP-based coatings in terms of functionality (self-healing properties, corrosion indicators, self-cleaning, anti-icing, antifouling, antireflection, etc.), considering the nature of specific metals and alloys to be protected along with economical and environmental factors.
- Automatic diagnosis of the metal state in several structures will be critical to their "smart" maintenance by CP-based coatings, which present a great potential for built-in coating condition monitoring.

Index

A

ABA, *see* 4-Aminobenzoic acid
ABF-G, *see* Aminobenzoyl group-functionalized graphene
ABSAs, *see* Aminobenzenesulfonic acids
ACAT, *see* Amine-capped aniline trimer
Activation polarization, 31
Adhesion promoter (AP), 81
AFM, *see* Atomic force microscopy
Alternating current electrochemical method (electrochemical impedance spectroscopy), 129–155
    equivalent circuit models, 146, 155
    experiment and data analysis, 133–155
    impedance spectrum for capacitor, 132
    Ohm's law, 129
    series resistance and capacitor, 132
    sinusoidal potential, 131
    theory, 129–132
Amine-capped aniline trimer (ACAT), 173, 178
Aminobenzenesulfonic acids (ABSAs), 169
4-Aminobenzoic acid (ABA), 94
Aminobenzoyl group-functionalized graphene (ABF-G), 94
Ammonium persulfate (APS), 62, 179
Aniline oligomers (AOs), 170
Anodic protection, 34, 39–44
Anticorrosion properties, characterization of, 109–158
    alternating current electrochemical method (electrochemical impedance spectroscopy), 129–155
    Bode plots, 134, 140, 154, 155
    charge transfer resistance, 133, 140
    equivalent circuit models, 146, 155
    experiment and data analysis, 133–155
    impedance analysis, 153
    impedance axis, 133, 136
    impedance spectrum for capacitor, 132
    ionic resistance of coating, 135
    Nyquist plot, 133, 138, 139
    Ohm's law, 129
    parallel resistance and capacitor, 132
    pore resistance, 135
    Randles circuit, 133, 134, 155
    series resistance and capacitor, 132
    sinusoidal potential, 131

    solution resistance, 133
    theory, 129–132
    uncompensated resistance, 133
    Warburg impedance, 135, 146
    basics of corrosion rate measurements, 109–110
    corrosion current density, 110
    Faraday's law, 109
    limitation, 110
    weight loss measurements, 110
    direct current electrochemical methods (half cell potential measurements), 111–128
    Butler-Volmer equation, 115
    comparative studies, 121
    corrosion rate measurement results, 118
    corrosion tendency, 111
    cyclic polarization, 124–126
    emeraldine base coated steel, 112
    extrapolation rules, 117
    Faraday constant, 115
    hysteresis loop, absence of, 125
    linear polarization resistance, 126–128
    open circuit potential, 111–115
    oxalic acid, 111, 120
    PAN–HCl-containing paint, 120
    polarization techniques, 115–128
    polyvinyl butyral, 114
    potentiodynamic measurements, 122–124
    sample notations, 119
    saturated calomel electrode, 113
    Stern Geary equation, 127
    Tafel extrapolation, 115–122
    zero-resistance ammetry, 156–157
    galvanic corrosion resistance, 156
    operational amplifier, 156
AOs, *see* Aniline oligomers
AP, *see* Adhesion promoter
APS, *see* Ammonium persulfate
Atomic force microscopy (AFM), 169, 181

B

Benzenesulfonate (BS), 79
Bilayer/multilayer coatings, 161
Blistering of coatings, 162
Bode plots, 133, 140, 154, 155
BS, *see* Benzenesulfonate
Butler-Volmer equation, 115

191

Milton Keynes UK
Ingram Content Group UK Ltd.
UKHW021622071024
449327UK00020BA/1156

9 780367 397661